NUMERICAL ANALYSIS—
The Mathematics of Computing
VOLUME 2

Volume 1 provides a lucid introduction to the subject whether the student be at school, college or university (It is suitable for the *first* year G.C.E. A-Level course in Numerical Analysis (Computions) and for the *whole* of the A-Level courses which include a limited amount of numerical analysis.)

NUMERICAL ANALYSIS—
The Mathematics of Computing
VOLUME 2

W. A. WATSON, B.Sc.

Head of Mathematics Department,
The Sweyne School, Rayleigh, Essex.

T. PHILIPSON, M.Sc.

Head of Mathematics Department,
Medway and Maidstone College of Technology,
Chatham, Kent.

P. J. OATES, B.Sc.Tech.

Head of Mathematics Department,
The Grove School,
Market Drayton, Shropshire.

EDWARD ARNOLD

© W. A. Watson, T. Philipson and P. J. Oates 1969

First published 1969
by Edward Arnold (Publishers) Ltd.
25 Hill Street, London W1X 8LL

Reprinted 1971, 1974

ISBN: 0 7131 2239 0

Printed in Great Britain by Page Bros (Norwich) Ltd.

Foreword

Mathematics plays an essential part in a wide variety of subjects. These subjects not only include all branches of Engineering and the Physical and Biological Sciences, but also such techniques as Business Management, Factory Operation, Traffic Control and Logic. Moreover, the further any of these subjects is studied, the more mathematical it is likely to become.

Mathematical studies involve the student in the solution of problem exercises so that the understanding of concepts, theories and their applications is deepened. Frequently the exercises involving numerical calculations are especially constructed so that the solutions are integers or simple rational fractions. If the exercises the student attempts are exclusively of this type it can result in the same mathematical methods, when used in the laboratory to calculate results from experiments, appearing to be very different and difficult. But in spite of the computational difficulties it is essential that the student should be able to complete calculations involving numbers which are not integers or simple rational fractions because it is calculations of these types which arise so frequently from actual real life situations.

Again at appropriate stages in his development, the student should be helped to realize that the logical, step by step sequence by which his mathematical tuition progresses (ideal though this sequence may be) only covers a limited range: problems being included only if formal mathematical solutions exist or can be improvised by exercising a little ingenuity. The solution of many problems, even quite elementary ones, which arise in engineering and science depends on equations which cannot be solved analytically (i.e. by using algebra and calculus to produce a formula), and yet it often happens on account of the practical, technological or economic significance of these problems that solutions must be provided and provided quickly and accurately. The way out of this apparently impossible dilemma is to utilize suitable numerical methods to provide acceptable numerical solutions for these problems: the solutions being acceptable if they can be stated to be of an accuracy which is satisfactory in the circumstances of the problem.

This indicates, rather briefly, why it is necessary to consider methods of obtaining numerical solutions to practical problems. But the reader will find that the usefulness of these methods is not restricted to those cases in which no analytical solutions can be found. Indeed there are many cases in which an analytical solution exists, but that as this is of a somewhat involved nature it is preferable to utilize a numerical method on account of the ease with which a numerical solution may thus be obtained.

Numerical Analysis—the Mathematics of Computing Volumes 1 and 2 attempt to provide an elementary and sound introduction to some of the most important methods of obtaining numerical solutions of known accuracy to a wide variety of problems.

The study of these numerical methods implies some prior knowledge of the mathematics involved. For example, before reading Chapter 2 a student would need to be aware of the need to obtain solutions of equations and have met some of the difficulties in solving equations by the usual direct methods. Again, before studying Chapter 4 a student would need to be familiar with the meaning of an integral and to have had some experience in evaluating definite integrals of some known elementary functions.

In addition to this general mathematical background essential for a full appreciation of certain chapters, use has also been made of a few important theorems in the derivation of some of the methods and formulae. To help the beginner brief notes on the binomial theorem and Taylor's theorem have been included in Volume 1 Chapter 1. It follows that the content of the present volumes is intended to be read concurrently with, or perhaps following, a Mathematics course of G.C.E. A. level or Ordinary National Certificate Standard.

Volume 2 contains: (a) a detailed discussion of interpolation; (b) methods. for numerical solution of differentiation and introductory work on the numerical solution of differential equations and (c) material on curve fitting by least squares and the summing of slowly convergent series.

Two of the authors have found that Volume 1 is very suitable for the first year of a two year G.C.E. A-level course in Numerical Analysis (Computations) and that this course is completed in the second year by using the greater part of the material in Volume 2. However some A-level courses which include Numerical Analysis do not include as much as this and for them Volume 1 will be a self-contained course in the subject.

The experience of the third author indicates that the two volumes will also be very suitable for a variety of courses at Ordinary National Certificate and Diploma, Higher National Certificate and Diploma and University levels.

Therefore there is reason to be confident that the volumes will be of real value to students and teachers in schools and further and higher educational establishments.

E. Kerr

Paisley
1968

Preface

From practical experience the authors believe that there exists a need for a text book on elementary numerical analysis, suitable for use as an introduction of the subject into the sixth forms of secondary schools and for courses in further education. With mathematical knowledge advancing so rapidly and numerical analysis an accepted part of mathematical degree courses, there is a case for the introduction of some, if not all, of the topics we have considered into the school curriculum. Hand-calculating machines are becoming readily available and, in the future, knowledge of how to use such mechanical aids will be essential to students and technologists. The study of numerical analysis also introduces the student to mathematical ideas and techniques which, as yet, are not generally incorporated into school syllabuses.

The authors wish to acknowledge the considerable help given to them in this venture by ADDO Ltd. It was under the auspices of ADDO Ltd. that they were first introduced to each other and since that day have been able to meet together at frequent intervals to discuss and amend draft material for the book.

The authors find it difficult to express in words the great debt they owe to Dr. Kerr, Principal of Paisley College of Technology, who has acted as general editor throughout the years of preparation. Dr. Kerr has given considerable help and advice most freely and generously.

Finally our grateful thanks to Mrs. Jill Pilkington (née Swindell) for all the secretarial help she has given in typing the manuscript.

We dedicate this book to all our students who have experienced our tuition, worked through the examples and in some cases contributed original examples themselves. The authors wish to thank the Associated Examining Board and The Cambridge Local Examinations Syndicate for permission to include questions set by them in past examination papers.

W.A.W.
T.P.
P.J.O.

1969

Contents

1

Interpolation Formulae

1.1 INTERPOLATION

When the method of Linear Interpolation was considered in Vol. I, Chapter 3, § 3.6.1 and Chapter 8, we saw how it was possible to calculate values of the function which lie between the tabulated values. In § 3.6.2 it was explained how linear interpolation, involving an evaluation along the chord joining two adjacent tabular values on the graph, necessarily introduced an error into the answer obtained. However, provided the adjacent points are very close, the error introduced will be small.

1.1.1 A linear interpolation formula can be established if we consider two adjacent entries in a table of differences, with values (x_0, f_0) and (x_1, f_1). Then these are the coordinates of two adjacent points of a polynomial curve passing through the tabular points given in the table of differences. [It is the establishment of such polynomial equations which forms the main part of this chapter].
In the fig. 1.1.1, $A(x_0, f_0)$; $B(x_1, f_1)$.

$$\text{gradient of chord AB} \quad = \frac{f_1 - f_0}{h}$$

$$= \frac{\Delta f_0}{h}$$

$$\therefore \text{ Equation of AB} \equiv y - f_0 = \frac{\Delta f_0}{h}(x - x_0)$$

i.e.
$$y = f_0 + \frac{\Delta f_0}{h}(x - x_0)$$

Thus using this result to find the value of the function at a point where $x = x_p$ and $x_p = x_0 + ph$ so that $0 < p < 1$.
We have

$$y = f_0 + \frac{\Delta f_0}{h}(x_p - x_0)$$

giving
$$y = f_0 + p \,.\, \Delta f_0$$

$$\text{or } f(x_p) = f_0 + p \,.\, \Delta f_0$$

which is a linear formula containing only first differences, used for linear interpolation.

This value $f(x_p)$ is represented by the point P on the diagram and hence there is an error represented by PQ. Later on in this chapter we shall show that this error is less than the permissible error in the interpolate of $\pm \frac{1}{2}$ in the last significant digit of the interpolate, if $|\Delta^2 f| < 4$.

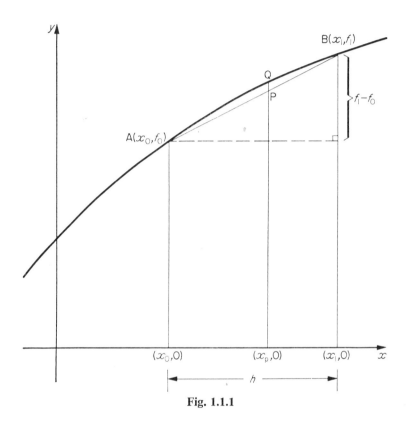

Fig. 1.1.1

1.1.2 Interpolation formulae or interpolation polynomials which we shall consider in this chapter are more general formulae, which include linear interpolation as a special case. To obtain these more suitable functions, we must first consider some elementary properties of 'symbolic difference operators', one of which, Δ, we have already met in Vol. I, Chapter 5, §.5.4. There the operator Δ, acting on the function value f_0, written Δf_0 implied the operation $f_1 - f_0$.

Similarly $\Delta f_{-2} = f_{-1} - f_{-2}$ etc.

We have much more to say about this and other operators later in the chapter. To help clarify the different notations, when these operators are introduced, the function $f(x) \equiv x^3$ for $x = 1(0 \cdot 1)1 \cdot 5$ is tabulated below.

x	$f(x) \equiv x^3$	1st Diff. $(\Delta f, \nabla f, \delta f)$	2nd Diff. $(\Delta^2 f, \nabla^2 f, \delta^2 f)$	3rd Diff. $(\Delta^3 f$ etc.$)$
x_{-2}	1	1 (f_{-2})		
			331	
x_{-1}	1·1	1·331 (f_{-1})		66
			397		6
x_0	1·2	1·728 (f_0)		72
			469		6
x_1	1·3	2·197 (f_1)		78
			547		6
x_2	1·4	2·744 (f_2)		84
			631	
x_3	1·5	3·375 (f_3)		

Note again, that for a cubic, with exact pivotal values, i.e. exact f_0, f_1, etc., the third order differences are constant.

1.2   THE FORWARD DIFFERENCE OPERATOR, Δ

In Vol. I, Chapter 5, § 5.4 we have Δ defined, by the relationship
$\Delta f(x) = f(x + h) - f(x)$ (where 'h' is the tabular interval)
Introducing a suffix notation we had
$\Delta f(x_n) = f(x_n + h) - f(x_n)$ which from now on we will write in a more concise and conventional form

$$\Delta f_n = f_{n+1} - f_n \tag{1}$$

Carrying this idea further, we used the operator on the first differences, etc., giving

$$\Delta(\Delta f_n) \equiv \Delta^2 f_n = \Delta f_{n+1} - \Delta f_n$$

and repeating the process r times gives

$$\Delta^r f_n = \Delta^{r-1} f_{n+1} - \Delta^{r-1} f_n \tag{2}$$

which is an *algebraic* relationship, involving the forward difference operator and relates the column containing the rth differences with the preceding column containing the $(r - 1)^{\text{th}}$ differences.

It is important to bear in mind, that whenever we are dealing with operators, they do not imply the actual numerical values obtained, which will naturally vary with different 'tables of values', but do imply the 'operation of differencing' involved.

1.3   THE SHIFT OPERATOR, E

This operator is used to represent movement from one position in a column of a 'table of differences' to the next position in the same column.

We define $E\{f(x)\} = f(x + h)$ (where 'h' is the tabular interval), i.e. we have moved down from one functional value to the next. Referring to the table of values for x^3, given in Section 1.1.2 for $f_0 = 1\cdot728$, $E(f_0) \equiv f_1 = 2\cdot197$, i.e. we have moved one place down the column. Repeating this process, and developing further the notation:

$E[E\{f(x)\}]$ implies $E[f(x + h)]$ which is $f(x + 2h)$ and then modifying the notation to write $E[E\{f(x)\}]$ as $E^2\{f(x)\}$ we have $E^2\{f(x)\} = f(x + 2h)$ and more generally

$$E^p\{f(x)\} = f(x + ph) \tag{3}$$

implying that we move p intervals of the tabular interval, downwards (forwards) if p is positive: similarly we move up (backwards) if p is negative.

Introducing a more concise suffix notation, (3) becomes

$$E^p(f_n) = f_{n+p}$$

from which, for example, we get,

$$E^2(f_0) = f_2 \; ; \; E^3(f_{-1}) = f_2 \quad \text{and} \quad E^{-1}(f_0) = f_{-1}.$$

Later, when we are considering central differences and associated formulae, we will need to use the shift operator for non-integral values. We consider $E^{\frac{1}{2}}(f_0)$ to mean $f_{\frac{1}{2}}$; i.e. $f(x_0 + \frac{1}{2}h)$ which is the value of the function at $x = x_0 + \frac{1}{2}h$, mid-way between two of the tabulated function values.

Referring to the table in section 1.1 $f_0 = (1\cdot2)^3$ then $E^{\frac{1}{2}}(f_0) = (1\cdot25)^3 = 1\cdot953\,125$, the function value mid-way between f_0 and f_1.

Similarly $E^{\frac{1}{2}}(f_{\frac{1}{2}}) = f_1$; $E^{\frac{1}{2}}(f_{3/2}) = f_2$; $E^{\frac{1}{2}}(f_2) = f_{5/2}$.

In the same way as for the Δ operator, we may apply the shift operator elsewhere in the difference table.

So $E^2(\Delta f_n) = \Delta f_{n+2}$ which implies that we have moved two places down the first difference column.

1.3.1 A relationship between the shift operator (E) and the forward difference operator (Δ) is now established.

From (1) we have $f_n + \Delta f_n = f_{n+1}$ and introducing the idea of an algebraic notation, we write

$$(1 + \Delta)f_n = f_{n+1}$$

also from (3)

$$E(f_n) = f_{n+1}$$

Equating the left hand side of these two results, we have

$$E(f_n) = (1 + \Delta)f_n$$

which in terms of the operators implies

$$E = 1 + \Delta \tag{4}$$

Note that these operators, in many respects behave like algebraic symbols and are used in similar ways.

For example, consider $E^2(f_n)$ and $(1 + \Delta)^2 f_n$, which from (4) should be equivalent.

We consider them in turn,

(a) $E^2(f_n) = f_{n+2}$

(b) $(1 + \Delta)^2 f_n$ gives $(1 + 2\Delta + \Delta^2)f_n$

i.e. $f_n + 2\Delta f_n + \Delta^2 f_n$

But, from the definition of Δ, we have $\Delta^2 f_n = \Delta f_{n+1} - \Delta f_n$ which, on substituting gives

$$(1 + \Delta)^2 f_n = f_n + 2\Delta f_n + (\Delta f_{n+1} - \Delta f_n)$$
$$= f_n + \Delta f_n + \Delta f_{n+1}$$

but $\qquad \Delta f_{n+1} = f_{n+2} - f_{n+1}$ and $\Delta f_n = f_{n+1} - f_n$ giving

$$(1 + \Delta)^2 f_n = f_n + (f_{n+1} - f_n) + (f_{n+2} - f_{n+1})$$

Hence $\qquad (1 + \Delta)^2 f_n = f_{n+2}$

which verifies algebraically that

$$E^2 = (1 + \Delta)^2$$

Using a similar method, it is possible to prove that

$$E^n = (1 + \Delta)^n \quad \text{(for positive integral } n)$$

1.4 THE GREGORY–NEWTON FORWARD DIFFERENCE INTERPOLATION FORMULA

Suppose that we have been given a series of values of an unknown function evaluated at equal intervals of the independent variable. The problem is to find a value of the function for a value of the variable lying between two of the given function values.

A graphical consideration of the problem will help. Consider a set of rectangular axes and let $x_1, x_2, x_3, x_4 \ldots x_n$ be 'n' values of the independent variable, at equal intervals and $y_1, y_2, y_3, y_4 \ldots y_n$ be the corresponding given function values.

In the figure 1.4, we plot only four points, i.e. consider the case, given only four function values.

There exist an infinity of smooth curves passing through the given points. (Only if special information is available, will the curve be unique.) The diagram shows two possible curves, each taking the function values at x_1, x_2, x_3, x_4 but having much different values between these points. Because the simplest functions are polynomials, we attempt to find, in each case if possible, a polynomial of degree 'n' passing through the given 'n' points. This polynomial will take on the given function values and we aim to obtain accuracy for non-tabular values of x and for this accuracy to be of a known order. Such a polynomial is called an Interpolation Formula. Once it is found, this polynomial does not yield a unique solution, because we may consider any suitable number of terms to give a function of any degree. This would give different degrees of the interpolation polynomial leading to different approximations for the interpolated value. Also any values obtained, lying between the given function values, must desirably have similar accuracy to that of the original data. This problem of accuracy is discussed later in the chapter.

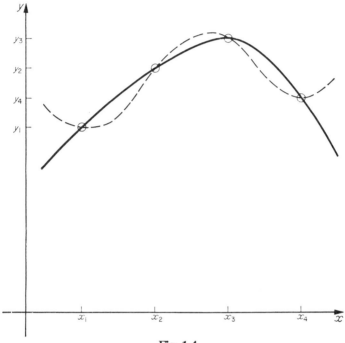

Fig. 1.4

We now attempt to find such a polynomial $f(p)$, such that its value f_p is the function value at the point x_p of the independent variable. The notation used is as follows:

$$f(p) \text{ denotes } f(x_p)$$

i.e. $f(p) \equiv f(x_p) = f(x_0 + ph)$ where x_0 is one of the given function values and $x_0 + ph$ (h the tabular interval) is a value between x_0 and x_1, with $0 < p < 1$. Then f_p the function value so calculated is the value we use as an approximation to the true value at x_p.

This polynomial, we are about to establish, uses the forward differences $\Delta f, \Delta^2 f, \ldots$ which follow the forward sloping line shown in Vol. I, Chapter 5, § 5.4.

from (3) we have $E^p(f_0) = f_p$ [where $f_p = f(x_0 + ph)$]

then using (4) $E^p(f_0) = (1 + \Delta)^p(f_0)$

[where p may be integral or fractional]. Equating the right hand sides gives

$$f_p = (1 + \Delta)^p f_0$$

and applying the binomial theorem we have

$$f_p = \left\{ 1 + \frac{p}{1!} \Delta + \frac{p(p-1)}{2!} \Delta^2 + \frac{p(p-1)(p-2)}{3!} \Delta^3 \ldots \right\} f_0$$

giving

$$f_p = f_0 + \frac{p}{1!} \Delta f_0 + \frac{p(p-1)}{2!} \Delta^2 f_0 + \frac{p(p-1)(p-2)}{3!} \Delta^3 f_0 + \ldots$$

which is known as the *Gregory–Newton forward difference interpolation formula or function*; discovered by James Gregory about 1670. The same formula was published by Newton in his 'Methodus Differentialis' 1711 which was known to have been written in the 1670s.

This formula will be used to interpolate between two adjacent values of the independent variable and these will be taken as x_0 and x_1, so that, as shown above for x_p, $0 < p < 1$. For values of p in this range the coefficients of the Gregory–Newton formula are decreasing in value.

Further if $0 < p < 1$ then the above interpolation formula will contain an infinite number of terms. Thus, if the tabular values are accurate, theoretically f_p can be evaluated to any required degree of accuracy, by taking a sufficient number of terms of the formula. The accuracy attainable in practice will be discussed later.

Hence, if the formula is truncated after $(n + 1)$ terms, we have the Gregory–Newton formula in the form

$$f_p = f_0 + \frac{p}{1!} \Delta f_0 + \frac{p(p-1)}{2!} \Delta^2 f_0 + \frac{p(p-1)(p-2)}{3!} \Delta^3 f_0 + \ldots.$$

$$+ \frac{p(p-1)(p-2)\ldots(p-n+1)}{n!} \Delta^n f_0 \qquad (5)$$

(Note how the first two terms of the right-hand side give the formula for linear interpolation).

We now show that (5) is a polynomial through the $(n + 1)$ points (x_0, f_0); (x_1, f_1); ... (x_n, f_n) and to do this we make use of theory given in Vol. I, Chapter 5, § 5.5 for finding the equation of a polynomial in terms of differences using factorial polynomials. We require an expansion for $f(x)$ if the function is *not* tabulated in unit intervals and *not* evaluated at $x = 0$, i.e. the most general case. From Vol. I, Chapter 5, §5.5.4 we have

$$f(x) = f(0) + \Delta f_0[x] + \frac{\Delta^2 f_0}{2!} [x]^2 + \ldots + \frac{\Delta^n (f_0) [x]^n}{n!}$$

which is a polynomial, using unit intervals.

For non-unit intervals and not evaluated at $x = 0$ we make the substitution

$$z = \frac{x - x_0}{h}$$ where x_0 is the first tabulated point and h is the interval.

giving $x = hz + x_0$ and $f(x) = f(hz + x_0)$ which gives $f(x) = F(z)$ say. Thus $F(0) = f(x_0)$; $F(1) = f(x_0 + h)$ or $f(x_1)$.

Hence
$$\Delta f(0) = F(1) - F(0)$$
$$= f(x_1) - f(x_0)$$

i.e.
$$\Delta F(0) = \Delta f(x_0)$$

similarly
$$\Delta^2 F(0) = \Delta^2 f(x_0) \text{ etc.}$$

But
$$F(z) = F(0) + \Delta F(0)\,[z] + \frac{\Delta^2 F(0)\,[z]^2}{2!} + \ldots + \frac{\Delta^n F(0)\,[z]^n}{n!}$$

Hence since $F(z) \equiv f(x)$ and $z = \dfrac{x - x_0}{h}$, substituting,

$$f(x) = f(x_0) + \Delta f(x_0)\left[\frac{x - x_0}{h}\right] + \frac{\Delta^2 f(x_0)}{2!}\left[\frac{x - x_0}{h}\right]^2 + \ldots$$

$$\ldots + \frac{\Delta^n f(x_0)}{n!}\left[\frac{x - x_0}{h}\right]^n$$

and expanding,

$$f(x) = f(x_0) + \Delta f(x_0) \cdot \frac{x - x_0}{h} + \frac{\Delta^2 f(x_0)}{2!}\frac{(x - x_0)(x - x_0 - h)}{h^2} + \ldots$$

$$\ldots + \frac{\Delta^n f(x_0)}{n!}\frac{(x - x_0)(x - x_0 - h)\ldots(x - x_0 - h(n-1))}{h^2}$$

which is the required formula for the case of non-unit intervals when the polynomial is not tabulated at $x = 0$.

We use this result to find the value of the function between the pivotal points, i.e. substituting $x = x_0 + ph$ we have

$$f(x_0 + ph) \text{ or } f_p = f(x_0) + \Delta f(x_0)\frac{x_0 + ph - x_0}{h} + \ldots + \frac{\Delta^2 f(x_0)}{2!}\frac{ph(ph - h)}{h^2}\ldots$$

$$\ldots + \frac{\Delta^n f(x_0)}{n!}\frac{(ph)(ph - h)\ldots(ph - (n-1)h)}{h^n}$$

which, on using the standard notation $f(x_0) = f_0$, and cancelling the h's, gives

$$f_p = f_0 + \Delta f_0 \cdot p + \frac{\Delta^2 f_0}{2!}p(p-1) + \ldots + \frac{\Delta^n f_0}{n!}p(p-1)(p-2)\ldots(p-n+1)$$

which is identical to (5) and is the polynomial through the given $(n + 1)$ points, as required. We have established that the operator method used in obtaining expression (5) does indeed yield a polynomial of degree n passing through a given set of $(n + 1)$ points. Later we shall use operator methods to produce other formulae and in each case it could be proved, by a similar method, that these are polynomials with the required property.

1.4.1 The restrictions of the Gregory–Newton forward difference interpolation formula in practice need noting. We have considered the Gregory–Newton polynomial first because it only required elementary properties and application of difference operators to establish it. Because the formula uses forward differences, its practical application is mainly restricted to finding functional values situated near the top of a table of differences, as we demonstrate below. If functional values are required elsewhere in the table, it is standard practice to use a central-difference interpolation polynomial, such as Bessel's or Everett's interpolation formula both of which are more advantageous than the Gregory–Newton formula. We deal with both these formulae later.

It is to be noted that the differences involved in this formula are obtained from the functional values which all lie to one side of x_0 and f_0, so that the higher order of differences are less and less related to the interval in which the interpolation is being carried out.

Worked example 1

Consider the following example, which makes use of the table of values given in Vol. I, Chapter 5, § 5.5.4 example 3.

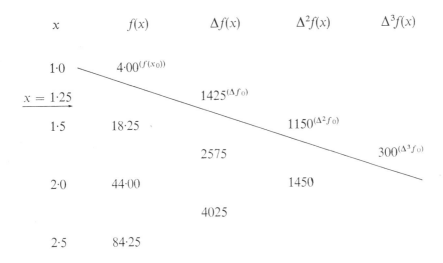

x	$f(x)$	$\Delta f(x)$	$\Delta^2 f(x)$	$\Delta^3 f(x)$
1·0	$4\cdot00^{(f(x_0))}$			
$x = 1\cdot25$		$1425^{(\Delta f_0)}$		
1·5	18·25		$1150^{(\Delta^2 f_0)}$	
		2575		$300^{(\Delta^3 f_0)}$
2·0	44·00		1450	
		4025		
2·5	84·25			

We evaluate $f(x)$ for $x = 1\cdot25$.
Since $\Delta^4 f(x)$ and higher-order differences are zero, we truncate the Gregory–Newton interpolation polynomial (5) to the first four terms and use

$$f_p = f_0 + p \cdot \Delta f_0 + \frac{p(p-1)}{2!} \Delta^2 f_0 + \frac{p(p-1)(p-2)}{3!} \Delta^3 f_0$$

We proceed as follows:

(i) First calculate the value of p, known as the interpolating factor.

In this example $x_0 = 1; x_p = 1\cdot25; h = 0\cdot5$

then using $\qquad\qquad x_p = x_0 + ph$

$$1\cdot25 = 1 + p.(0\cdot5)$$

$$\Rightarrow p = 0\cdot5$$

(ii) Substituting for $p = 0\cdot5$, etc., we have

$$f_p \text{ or } f(1\cdot25) = 4 + (0\cdot5)(14\cdot25) + \frac{(0\cdot5)(-0\cdot5)}{2!}(11\cdot50) +$$

$$\frac{(0\cdot5)(-0\cdot5)(-1\cdot5)}{3!}(3\cdot00)$$

For this we introduce a column method of presentation as the most convenient method of calculating the coefficients.

	$\Delta^r f_0$			Coefficient
f_0	4·00	1	1	1
Δf_0	14·25	p	0·5	0·5
$\Delta^2 f_0$	11·50	$p \times \dfrac{p-1}{2}$	$0\cdot5 \times \dfrac{-0\cdot5}{2}$	−0·125
$\Delta^3 f_0$	3·00	$\dfrac{p(p-1)}{2} \times \dfrac{p-2}{3}$	$-0\cdot125 \times \dfrac{-1\cdot5}{3}$	0·0625

As we have seen, the coefficients in this formula are binomial coefficients. The third and fourth columns above show how they may be calculated as a continuous process. Many tables are available from which these values can be read. In this case, however, we evaluate the coefficients and then evaluate f_p as a continuous process (Σxy) on the machine.

i.e. evaluate $\quad f_p = (4).(1) + (14\cdot25).(0\cdot5) + (11\cdot50)(-0\cdot125) + (3\cdot00)(0\cdot0625)$

giving $\qquad f_p = 9\cdot875$

Since in Vol. I. Chapter 5, § 5.5.4, ex. 3, we established the function as $f(x) \equiv 4x^3 + 5x^2 - 3x - 2$ and evaluating $f(1\cdot25)$ by 'Nested multiplication' gives $f(1\cdot25) = 9\cdot875$, we have an exact value because the functional values of the table were exact and we used an interpolation polynomial of sufficiently high order.

Worked example 2

From Chambers Six Figure Mathematical Tables for the Gamma Function $\Gamma(x)$ when $\Gamma(x) = \int_0^\infty e^{-t}\, t^{x-1}\, dt$ we have the following values

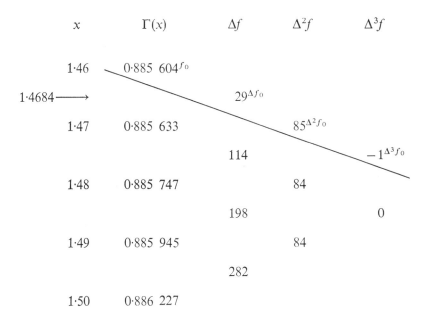

x	$\Gamma(x)$	Δf	$\Delta^2 f$	$\Delta^3 f$
1·46	0·885 604^{f_0}			
1·4684 $\longrightarrow$		29$^{\Delta f_0}$		
1·47	0·885 633		85$^{\Delta^2 f_0}$	
		114		$-1^{\Delta^3 f_0}$
1·48	0·885 747		84	
		198		0
1·49	0·885 945		84	
		282		
1·50	0·886 227			

We require to find $\Gamma(x)$ when $x = 1\cdot4684$.

We observe that not all the third-order differences are zero or constant.

As before, we first calculate the value of the interpolating factor p.

$$x_0 = 1\cdot46;\ x_p = 1\cdot4684;\ h = 0\cdot01.$$

using
$$x_p = x_0 + ph$$

$$\therefore p = \frac{x_p - x_0}{h}$$

$$= \frac{0\cdot0084}{0\cdot01}$$

$$\Rightarrow p = 0\cdot84$$

Further, if we apply the Gregory–Newton interpolation function, as the coefficient of the term containing the third-order difference is $p(p - 1)(p - 2)/3!$ which when $p = 0\cdot84$ is $\simeq 0\cdot03$, this combined with the third-order difference is negligible, and we may truncate the interpolation formula to

$$f_p = f_0 + p.\,\Delta f_0 + \frac{p(p - 1)}{2!}\Delta^2 f_0$$

	$\Delta^r f_0$		Coefficient
f_0	0·885 604	1	1
Δf_0	0·000 029	p	0·84
$\Delta^2 f_0$	0·000 085	$\dfrac{p(p-1)}{2}$	—0·0672

Thus, the calculation we have to evaluate is

$$f_p = 0·885\ 604(1) + (0·84)\,(0·000\ 029) + (-\ 0·0672)\,(0·000\ 085)$$

and this can be calculated as a continuous process on the machine. By inspection we see that the Acc. will require a setting for 10 dec. pl. The decimal-point settings will be S.R. 6; C.R. 4 $\rightarrow$ Acc. 10.

Then the work on the machine proceeds as follows. The values taken from the differences are set on S.R. each time and once the decimal point markers are set initially they will not require altering until the evaluation is completed.

	S.R. (6)	C.R.(4)	Acc. (10)
Set 0·885 604 in S.R. and add into Acc.	0·885 604	1·0000	0·885 604 000 0
Clear S.R. and C.R.	0	0	0·885 604 000 0
Set 0·000 029 in S.R. and add by multiplying by 1·0000	0·000 029	1·0000	0·885 633 000 0
(Check that the Acc. now gives f_1). Adjust the C.R. to read 0·8400	0·000 029	0·8400	0·885 628 360 0*
Clear S.R. and C.R.	0	0	0·885 628 360 0
Set 0·000 085 in S.R. and SUB-TRACT with 0·0672 in C.R.	0·000 085	0·0672	0·885 622 648 0

Hence we have $f_p = 0·885\ 623$ (6D)

As we have said we cannot expect an answer more accurate than the initial data, so we have rounded the final result to 6D as the most accurate answer we can expect. This problem of accuracy is discussed later.

* It is of interest to note that, at this stage marked in the calculations, the Acc. now shows the value of $f(1·4684)$ had we used linear interpolation and gives 0·885 628. Comparing this result with the final one obtained above, we have an example of the inaccuracy of linear interpolation, even though the tabular interval is small.

1.5 THE BACKWARD DIFFERENCE OPERATOR, ∇

We define this operator by the relationship
$$\nabla f(x) = f(x) - f(x - h)$$
and using the suffix notation as in Vol. I, § 8.2 we have
$$\nabla f(x_n) = f(x_n) - f(x_n - h)$$
which still conforms to the general rule of establishing 1st order differences by subtracting one tabular or pivotal value from the entry immediately below. Thus using a notation where $x_n - h$ is x_{n-1} we have
$$\nabla f_n = f_n - f_{n-1} \tag{6}$$
and this gives $\qquad \nabla f_0 = f_0 - f_{-1}; \quad \nabla f_1 = f_1 - f_0; \ldots$

Similarly, as in Vol. I, §5.4, we may apply this ∇ operator on the column of first differences to give the second differences.

i.e. $\nabla(\nabla f_n) = \nabla(f_n - f_{n-1})$ using (6)

giving $\qquad \nabla^2 f_n = \nabla f_n - \nabla f_{n-1}$

so that $\qquad \nabla^2 f_{-1} = \nabla f_{-1} - \nabla f_{-2}; \quad \nabla^2 f_0 = \nabla f_0 - \nabla f_{-1}$, etc.

and continuing this process, we have,
$$\nabla^r f_n = \nabla^{r-1} f_n - \nabla^{r-1} f_{n-1} \tag{7}$$

Using this notation we may establish the normal table of differences,

		1st. diff.	2nd diff.	3rd. diff.	4th. diff.
x_{-2}	f_{-2}				
		∇f_{-1}			
x_{-1}	f_{-1}		$\nabla^2 f_0$		
		∇f_0		$\nabla^3 f_1$	
x_0	f_0		$\nabla^2 f_1$		$\nabla^4 f_2$
		∇f_1		$\nabla^3 f_2$	
x_1	f_1		$\nabla^2 f_2$		$\nabla^4 f_3$
		∇f_2		$\nabla^3 f_3$	
x_2	f_2		$\nabla^2 f_3$		
		∇f_3			
x_3	f_3				

This time we see that the differences with the same suffix lie on a backward sloping line.

Comparing this with the table of values in § 1.1.2 for $f(x) \equiv x^3$ we have the entry 72 in the 2nd differences corresponding to $\nabla^2 f_1$ in the above table.

We now establish a relationship between the backward difference operator, ∇, and the shift-operator, E.

By definition, we have

$$\nabla f_n = f_n - f_{n-1}$$

and

$$E^{-1} f_n = f_{n-1}$$

Thus we have

$$f_{n-1} = f_n - \nabla f_n$$

i.e.

$$E^{-1} f_n = f_n - \nabla f_n$$

$$E^{-1} f_n = f_n (1 - \nabla)$$

giving

$$E^{-1} = 1 - \nabla$$

or

$$E = (1 - \nabla)^{-1} \qquad (8)$$

1.6 THE GREGORY–NEWTON BACKWARD DIFFERENCE INTERPOLATION FORMULA

As before, we now obtain an interpolation formula which will enable us to evaluate the function for a value of the argument lying between pivotal values. As in 1.4

$$f_p = E^p(f_0)$$

substituting from (8) for E gives

$$f_p = (1 - \nabla)^{-p} f_0$$

Expanding the right-hand side using the binomial series, we obtain

$$f_p = \left\{ 1 - p(-\nabla) + \frac{-p(-p-1)}{2!} (-\nabla)^2 + \ldots \right\} f_0$$

giving

$$f_p = \left\{ 1 + p\nabla + \frac{p(p+1)}{2!} \nabla^2 + \frac{p(p+1)(p+2)}{3!} \nabla^3 + \ldots \right\} f_0$$

Thus

$$f_p = f_0 + p.\nabla f_0 + \frac{p(p+1)}{2!} \nabla^2 f_0 + \frac{p(p+1)(p+2)}{3!} \nabla^3 f_0 + \ldots \qquad (9)$$

which is the *Gregory–Newton backward difference interpolation formula*.

1.6.1 The restrictions of the Gregory–Newton backward difference interpolation formula need noting, as with the corresponding forward difference formula. It is generally only used when the functional value required is at the bottom end of the table of differences.

Note again that all the differences involved are on one side of the interval with which the calculation is concerned.

Worked example

We will make use of the table of values given in section 1.1.2 for the function $f(x) \equiv x^3$ and evaluate $f(1 \cdot 457)$. Then using

$$f_0 = 3 \cdot 375; \, \nabla f_0 = 0 \cdot 631; \, \nabla^2 f_0 = 0 \cdot 084; \, \nabla^3 f_0 = 0 \cdot 006$$

from
$$x_p = x_0 + ph$$
$$1 \cdot 457 = 1 \cdot 5 + ph$$
$$- 0 \cdot 043 = 0 \cdot 1 p$$
$$\Rightarrow p = - 0 \cdot 43$$

We use the first four terms of formula (9).
Tabulating the coefficients, as in previous examples:

	$\nabla^r f_0$		Coefficient
f_0	3·375	1	1
∇f_0	0·631	p	− 0·43
$\nabla^2 f_0$	0·084	$p \times \dfrac{p+1}{2}$	− 0·122 55
$\nabla^3 f_0$	0·006	$\dfrac{p(p+1)}{2} \times \dfrac{p+2}{3}$	− 0·064 134 5

Hence multiplying the second and fourth columns together in each row and adding the series of products is in fact evaluating

$$f_p = 3 \cdot 375(1) + (- 0 \cdot 43)(0 \cdot 631) + (- 0 \cdot 122\ 55)(0 \cdot 084) +$$
$$(- 0 \cdot 064\ 134\ 5)(0 \cdot 006)$$

By inspection, setting the decimal point markers
$$\text{S.R. 7; C.R. 3} \rightarrow \text{Acc. 10}$$
gives
$$f(1 \cdot 457) = 3 \cdot 092\ 990\ 993 \text{ which is } (1 \cdot 457)^3 \text{ exactly.}$$

Again we have an exact result because the function values were exact and we used the appropriate number of terms of the truncated polynomial interpolation formula. Further, although the function values were only given to 3D we retained 7D in the coefficients whilst multiplying to 10D. Normally it would not be worthwhile retaining so many decimal places. The student must always consider the problem of how many figures to retain in the interpolation coefficients. In general, one must work to a sensible number of decimal places, in keeping with the degree of accuracy possible. Later, when this problem is discussed more fully, we state that a sensible rule is, that the number of decimals in the coefficient should not be less than the significant figures in the difference. Thus, in the above example we could have worked with coefficients to 3D and rounded off to 3.093 (3D) with satisfactory accuracy. We only maintained this high accuracy because of the special circumstances.

1.7   EXAMPLES

1. (a) Using the relationship $E = \Delta + 1$ obtain an expression for the nth order difference $(\Delta^n f)$ of a function in terms of successive functional values.
(b) Use the formula obtained to evaluate $\Delta^3 f_0$ from the following functional values:

f_0	f_1	f_2	f_3	
1	1·331	1·728	2·197	(Check answer in Table of Differences in section 1.1.2)

2. (a) Using the relationship $E = 1 + \Delta$ obtain the Gregory–Newton formula for $E^x(f_0)$, where $x_0 \equiv 0$, in the form:

$$f(x) = f_0 + x\Delta f_0 + \frac{x(x-1)}{2!}\Delta^2 f_0 + \frac{x(x-1)(x-2)}{3!}\Delta^3 f_0 + ..$$

(b) A certain polynomial is tabulated at successive unit values as shown:

$x =$	0	1	2	3	4	5
$f(x) =$	0	−1	0	9	32	75

Establish the table of differences and substitute in the formula of part (a) to verify that the polynomial considered was

$$f(x) \equiv x^3 - 2x^2$$

3. For a certain polynomial we have the following table of values:

x	−2	−1	0	1	2	3
$f(x)$	−3	−19	−21	−21	−7	57

Use the method of Question 2 to establish the polynomial explicitly in terms of x.

4. Taken from Chamber's Six Figure Tables the values of e^x are as follows:

x	3·500	3·505	3·510	3·515	3·520
e^x	33·115 45	33·281 44	33·448 27	33·615 93	33·784 43

Evaluate $e^{3·5023}$ to 5D.

5. The probability integral $a = \sqrt{\dfrac{2}{\pi}} \displaystyle\int_0^x e^{-\frac{1}{2}t^2}\, dt$ has the following values

x	1·00	1·05	1·10	1·15	1·20	1·25
a	0·682 689	0·706 282	0·728 668	0·749 856	0·769 861	0·788 700

Evaluate a for $x = 1·235$

6. Show that $E = 1 + \Delta$ and $\Delta = \nabla(1 - \nabla)^{-1}$. Hence deduce that $1 + \Delta = (E - 1)\nabla^{-1}$. (Cambridge).

7. Using the Newton–Gregory backward difference interpolation formula evaluate $\sin 24° \, 47' \, 30''$ from the following set of values:

$\sin 24°$	$\sin 24° \, 10'$	$\sin 24° \, 20'$	$\sin 24° \, 30'$
0·406 737	0·409 392	0·412 045	0·414 693

$\sin 24° \, 40'$	$\sin 24° \, 50'$
0·417 338	0·419 980

1.8 THE CENTRAL DIFFERENCE OPERATOR, δ

In this section we introduce a third operator, the central difference operator defined by

$$\delta f_{\frac{1}{2}} = f_1 - f_0$$

or more generally

$$\delta f_{n+\frac{1}{2}} = f_{n+1} - f_n \text{ for all integral } n \qquad (10)$$

Note how the suffix of the left-hand side is the mean of the two suffices on the right-hand side.

Further the operator may be applied to any column of the difference table. The notation is as follows

$$\delta(\delta f) \text{ written } \delta^2 f$$

and

$$\delta^2(f_n) = \delta(\delta f_n)$$

$$= \delta(f_{n+\frac{1}{2}} - f_{n-\frac{1}{2}})$$

giving

$$\delta^2 f_n = \delta f_{n+\frac{1}{2}} - \delta f_{n-\frac{1}{2}}$$

In the table of differences given below, which has been established using the central difference operator, the whole notation follows these definitions.

In particular, notice that all the differences with the same suffix as the pivotal value will be along the same row and all occur as even order differences. That is, f_0, $\delta^2 f_0$, $\delta^4 f_0$ . . ., etc., lie along the same row. Also odd order differences lie on the line between f_n and f_{n+1} and have suffix $(n + \frac{1}{2})$. i.e. $\delta f_{\frac{1}{2}}$, $\delta^3 f_{\frac{1}{2}}$, $\delta^5 f_{\frac{1}{2}}$ . . . lie on same row.

(Remember that numerically the 'table of differences' is still calculated in the standard way).

		1st Diff.	2nd Diff.	3rd Diff.	4th Diff.	
		(δf)	$(\delta^2 f)$	$(\delta^3 f)$	$(\delta^4 f)$	
x_{-2}	f_{-2}					
		$\delta f_{-\frac{3}{2}}$				
x_{-1}	f_{-1}		$\delta^2 f_{-1}$			
		$\delta f_{-\frac{1}{2}}$			$\delta^3 f_{-\frac{1}{2}}$	
x_0	f_0		$\delta^2 f_0$			$\delta^4 f_0$
		$\delta f_{\frac{1}{2}}$		$\delta^3 f_{\frac{1}{2}}$		$\delta^5 f_{\frac{1}{2}}$
x_1	f_1		$\delta^2 f_1$		$\delta^4 f_1$	
		$\delta f_{\frac{3}{2}}$		$\delta^3 f_{\frac{3}{2}}$		
x_2	f_2		$\delta^2 f_2$			
		$\delta f_{\frac{5}{2}}$				
x_3	f_3					

1.8.1   A relationship between δ and E is established by considering a result of §1.8,

$$\delta f_{n+\frac{1}{2}} = f_{n+1} - f_n$$

i.e.
$$\delta f_{n+\frac{1}{2}} = E^{\frac{1}{2}} f_{n+\frac{1}{2}} - E^{-\frac{1}{2}} f_{n+\frac{1}{2}}$$

giving
$$\delta f_{n+\frac{1}{2}} = (E^{\frac{1}{2}} - E^{-\frac{1}{2}}) f_{n+\frac{1}{2}}$$

i.e.
$$\delta = E^{\frac{1}{2}} - E^{-\frac{1}{2}} \tag{11}$$

1.9   CENTRAL DIFFERENCE INTERPOLATION FORMULA—BESSEL'S

In the application of the previous two interpolation formulae the information which is used all lies on one side or the other of the interpolation interval. However, it seems intuitively obvious that better results would be obtained if the information used in the calculations is symmetrically placed about the interval in which the interpolation is required We are led to consider a formula involving central differences, of the form

$$f_p = f_0 + B_1\, \delta f_{\frac{1}{2}} + B_2\,(\delta^2 f_0 + \delta^2 f_1) + B_3\, \delta^3 f_{\frac{1}{2}} +$$

$$B_4\,(\delta^4 f_0 + \delta^4 f_1) + \dots \tag{12}$$

where the coefficients $B_1, B_2, B_3, \dots$ are called the Bessel interpolation coefficients.

Note that this suggested formula involves the use of the differences which lie along and between the rows across, from f_0 and f_1, i.e.

f_0		$\delta^2 f_0$		$\delta^4 f_0$ —	—	—	—	—	—	—	—
	$\delta f_{\frac{1}{2}}$		$\delta^3 f_{\frac{1}{2}}$		$\delta^5 f_{\frac{1}{2}}$ —	—	—	—	—	—	—
f_1		$\delta^2 f_1$		$\delta^4 f_1$ —	—	—	—	—	—	—	—

As before, the coefficients B_1, B_2, B_3 ... are functions of the interpolating fraction (p) of the interval over which the interpolation is to be carried out. We must find the form of these coefficients. There are various ways of doing this and we choose the method of expanding both sides of (12) in terms of Δ and then compare coefficients. To do this we need various auxiliary formulae and relationships connecting the difference operators.

From (11) we have $\delta = E^{\frac{1}{2}} - E^{-\frac{1}{2}}$

$$\therefore \quad E^{\frac{1}{2}}\delta = E^{\frac{1}{2}}(E^{\frac{1}{2}} - E^{-\frac{1}{2}})$$

$$\Rightarrow \quad E^{\frac{1}{2}}\delta = E - 1$$

But from (4) $\qquad E = 1 + \Delta$

$$\therefore \quad E^{\frac{1}{2}}\delta = \Delta \tag{13}$$

squaring gives $\qquad E\delta^2 = \Delta^2$

$$\Rightarrow \quad \delta^2 = \frac{\Delta^2}{1 + \Delta} \tag{14}$$

Using these relationships, we expand both sides of the formula (12) as functions of Δ. Then equating coefficients we can evaluate B_1, B_2, B_3, etc., as functions of 'p'.

For

$$E^p(f_0) = f_0 + B_1\delta(E^{\frac{1}{2}}f_0) + B_2(\delta^2 + \delta^2 E)f_0 + B_3\delta^3 E^{\frac{1}{2}}(f_0) + B_4(\delta^4 + \delta^4 E)f_0 \ldots$$

i.e.

$$E^p f_0 = (1 + B_1\delta E^{\frac{1}{2}} + B_2(\delta^2 + \delta^2 E) + B_3\delta^3 E^{\frac{1}{2}} + B_4(\delta^4 + \delta^4 E) + \ldots)f_0$$

giving

$$E^p = 1 + B_1\delta E^{\frac{1}{2}} + B_2(\delta^2 + \delta^2 E) + B_3\delta^3 E^{\frac{1}{2}} + B_4(\delta^4 + \delta^4 E) + \ldots$$

Now making the appropriate substitutions for E, $\delta E^{\frac{1}{2}}$, δ^2, in terms of Δ we have

$$(1 + \Delta)^p = 1 + B_1\Delta + B_2\left(\frac{\Delta^2}{1 + \Delta} + \Delta^2\right) + B_3\left(\frac{\Delta^2}{1 + \Delta} \cdot \Delta\right) +$$

$$B_4\left(\frac{\Delta^4}{(1 + \Delta)^2} + \frac{\Delta^4}{1 + \Delta}\right) + \ldots$$

Using the binomial series on both sides, left-hand side becomes

$$1 + p \cdot \Delta + \frac{p(p-1)}{2!}\Delta^2 + \frac{p(p-1)(p-2)}{3!}(\Delta^3) +$$

$$\frac{p(p-1)(p-2)(p-3)}{4!}(\Delta^4) \ldots$$

and the right-hand side becomes

$$1 + B_1\Delta + B_2(\Delta^2 - \Delta^3 + \Delta^4 - + \ldots + \Delta^2) +$$
$$B_3\Delta^3(1 - \Delta + \Delta^2 - \Delta^3 + \Delta^4 - + \ldots) +$$
$$B_4\Delta^4(1 - 2\Delta + 3\Delta^2 - 4\Delta^3 + - \ldots + 1 - \Delta + \Delta^2 - \Delta^3 + - \ldots) \ldots$$

Equating the coefficients:

$$B_1 = p$$

Coefficients of Δ^2: $\qquad\qquad 2B_2 = \dfrac{p(p-1)}{2}$

Coefficients of Δ^3: $\qquad B_2 + B_3 = \dfrac{p(p-1)(p-2)}{3!}$

Coefficients of Δ^4: $\quad B_2 - B_3 + 2B_4 = \dfrac{p(p-1)(p-2)(p-3)}{4!}$,

etc.

$\qquad$ this gives $B_2 = \dfrac{p(p-1)}{4}$

$$\therefore\ B_3 = \frac{p(p-1)}{4} + \frac{p(p-1)(p-2)}{3!}$$

$$= p(p-1)\left[\tfrac{1}{4} + \frac{p-2}{6}\right]$$

$$B_3 = \frac{p(p-1)(p-\tfrac{1}{2})}{3!}$$

$$\therefore\ 2B_4 = \frac{p(p-1)(p-2)(p-3)}{4!} + \frac{p(p-1)(p-\tfrac{1}{2})}{3!} - \frac{p(p-1)}{4}$$

$$= \frac{p(p-1)}{4!}[(p-2)(p-3) + 4(p-\tfrac{1}{2}) - 6]$$

$$\therefore\ 2B_4 = \frac{p(p-1)}{4!}[p^2 - 5p + 6 + 4p - 2 - 6]$$

$$= \frac{p(p-1)(p+1)(p-2)}{4!}$$

giving $B_4 = \dfrac{(p+1)(p)(p-1)(p-2)}{2.4!}$

Hence we have Bessel's interpolation formula or polynomial:

$$f_p = f_0 + p \cdot \delta f_{\frac{1}{2}} + \frac{p(p-1)}{4}(\delta^2 f_0 + \delta^2 f_1) + \frac{p(p-1)(p-\tfrac{1}{2})}{3!}\delta^3 f_{\frac{1}{2}} +$$

$$\frac{(p+1)p(p-1)(p-2)}{2.4!}(\delta^4 f_0 + \delta^4 f_1) + \ldots \qquad (15)$$

which has an infinite number of terms for $0 < p < 1$.

Note, that for any particular p in this range these Bessel coefficients decrease more rapidly than those of the Gregory–Newton forward and backward difference formulae. Thus fewer terms of the formula are needed to provide a result to the same degree of accuracy.

1.9.1 Worked example to consider the application of the above formula. The following function values are exact values of an unknown polynomial.

x	0·1	0·2	0·3	0·4	0·5	0·6
$f(x)$	1·172	1·008	0·878	0·782	0·720	0·692

Evaluate $f(x)$ for $x = 0.36$

	x	$f(x)$	δf	$\delta^2 f$
f_{-2}	0·1	1·172		
			-164	
f_{-1}	0·2	1·008		34
			-130	
f_0	0·3	0·878		34
			-96	
f_1	0·4	0·782		34
			-62	
f_2	0·5	0·720		34
			-28	
f_3	0·6	0·692		

To evaluate $f(0.36)$ we use the Bessel's interpolation polynomial truncated after the third term because $\delta^3 f_{\frac{1}{2}} = 0$

i.e. $$f_p = f_0 + p \cdot \delta f_{\frac{1}{2}} + \frac{p(p-1)}{4}(\delta^2 f_0 + \delta^2 f_1)$$

It is seen from the table of differences that the differences involved in this formula are obtained from functional values which lie on both sides of the interval within which the interpolation is required, as shown by the 'fan' in the table of values.

(i) To find the interpolating factor p

$$\text{using } x_p = x_0 + ph$$

$$\Rightarrow p = \frac{x_p - x_0}{h}$$

$$= \frac{0{\cdot}36 - 0{\cdot}3}{0{\cdot}1}$$

$$\text{gives } p = 0{\cdot}6$$

(ii) Evaluating the coefficients

			Coefficients
f_0	$0{\cdot}878$	1	1
$\delta f_{\frac{1}{2}}$	$-0{\cdot}096$	p	$0{\cdot}6$
$\delta^2 f_0 + \delta^2 f_1$	$0{\cdot}068$	$p \times \dfrac{p-1}{4}$	$-0{\cdot}06$

then multiplying the associated quantities from columns 2 and 4 and adding gives the required evaluation.

$$f(0{\cdot}36) = 0{\cdot}878 + (0{\cdot}6)(-0{\cdot}096) + (-0{\cdot}06)(0{\cdot}068)$$

which, working on a machine gives

$$f(0{\cdot}36) = 0{\cdot}816\ 32$$

Note, that just as in § 1.6, because of the exact values of the function values, the differences and the coefficients, the value of $f(0{\cdot}36)$ obtained is exact. Again, in these special circumstances, we see the value of retaining more decimal figures than that given in the original data. You may check the accuracy of the above result because the 'unknown' function giving the function values was

$$f(x) \equiv 1{\cdot}7x^2 - 2{\cdot}15x + 1{\cdot}37.$$

1.9.2 Interpolation and Allied Tables, prepared by H.M. Nautical Almanac Office and published by H.M.S.O., are a most important aid to computations. From now on we shall refer to the tables as *I.A.T.* Extensive use of these will be made in the remainder of the book.

The above formula (15) can be found on page 56 of *I.A.T.* but given in the form

$$f_p = f_0 + p\delta_{\frac{1}{2}} + B_2(\delta_0^2 + \delta_1^2) + B_3\delta_{\frac{1}{2}}^3 + \ldots$$

with the coefficients $B_2, B_3, \ldots$ as functions of 'p', given elsewhere on the same page. Note the concise notation which substitutes $\delta_{\frac{1}{2}}$ for $\delta f_{\frac{1}{2}}$, etc. However we leave further applications of *I.A.T.* until we reach section 1.11. where we use them to facilitate interpolation calculations.

1.9.3 Worked example in which the functional values of a polynomial are all rounded off to 5D.
Using the following finite difference table evaluate $f(0.243)$

x	$f(x)$	δf	$\delta^2 f$	$\delta^3 f$
0.23	2.952 81			
		1990		
0.24	2.972 71		67	
		2057		3
0.25	2.993 28		70	
		2127		0
0.26	3.014 55		70	
		2197		2
0.27	3.036 52		72	
		2269		
0.28	3.059 21			

We see, that this time, there is not a constant numerical value for the third difference column. This is because the functional values have been rounded off.
To find p, the interpolating factor

$$x_p = x_0 + ph$$
$$0.243 = 0.24 + p.(0.01)$$
giving $p = 0.3$ where $x_0 = 0.24$ and $h = 0.01$

Using Bessel's interpolation polynomial:

$$f_p = f_0 + p\delta f_{\frac{1}{2}} + \frac{p(p-1)}{4}(\delta^2 f_0 + \delta^2 f_1) + \frac{p(p-1)(p-\frac{1}{2})}{3!}\delta^3 f_{\frac{1}{2}}$$

(and no further terms required.)

Evaluating the coefficients:

		Coefficient		
1	1	1	f_0	2.972 71
p	0.3	0.3	$\delta f_{\frac{1}{2}}$	0.020 57
$p \times \dfrac{p-1}{4}$	$\dfrac{0.3 \times (-0.7)}{4}$	-0.0525	$\delta^2 f_0 + \delta^2 f_1$	0.001 37
$p \times \dfrac{p-1}{4} \times \dfrac{2(p-\frac{1}{2})}{3}$	-0.0175			
$\dfrac{p(p-1)(p-\frac{1}{2})}{3!}$	$\dfrac{-0.0525 \times -0.4}{3}$	$+0.007$	$\delta^3 f_{\frac{1}{2}}$	0.000 03

Multiplying the associated terms gives the evaluation

$$f(0 \cdot 243) = 2 \cdot 972\ 71 + (0 \cdot 3)\ (0 \cdot 020\ 57) + (-\ 0 \cdot 052\ 5)\ (0 \cdot 001\ 37) +$$
$$(0 \cdot 007)\ (0 \cdot 000\ 03)$$

and working to 9D in the Acc. gives

$$f_p = 2 \cdot 978\ 809\ 285$$

so that $f(0 \cdot 243) = 2 \cdot 978\ 81$ (5D) which is correct to 5D as can be checked against the answer $2 \cdot 978\ 806\ 750\ 3$, for the polynomial considered was $2 \cdot 33x^3 + 1 \cdot 71x^2 + 0 \cdot 8x + 2 \cdot 65$.

The above example shows clearly the effect that rounding off the functional values will produce, both on the final column of the 'table of differences' and also on the computed result.

By inspection, the final term involving $\delta^3 f_{\frac{1}{2}}$, gives $0 \cdot 000\ 000\ 21$ and could safely have been omitted without affecting the final result. As one would expect, if there are rounding-off errors in the functional values, the degree of accuracy of the final result could not be greater than that of the original data. Our result was rounded off to 6S.

1.10   EVERETT'S INTERPOLATION FORMULA

This is another central difference formula, with the differences used lying adjacent to the same lines as for Bessel's formula, but no odd-order differences are needed. Everett's formula would be most convenient to use if only even order differences are tabulated. Also, it can be used to check the accuracy of a calculation performed using Bessel's formula.

It is possible to derive Everett's formula in exactly the same way as we did Bessel's but we prefer to derive it directly from Bessel's formula, in order to demonstrate the direct relation between the two.

For *Everett's interpolation formula*, using Bessel's formula as in § 1.9, from (15) we have:

$$f_p = f_0 + p\delta f_{\frac{1}{2}} + \frac{p(p-1)}{4}(\delta^2 f_0 + \delta^2 f_1) + \frac{p(p-1)(p-\frac{1}{2})}{3!}\delta^3 f_{\frac{1}{2}} + \ldots$$

also by definitions for central difference notation we have:

$$f_1 - f_0 = \delta f_{\frac{1}{2}}; \ \delta^3 f_{\frac{1}{2}} = \delta^2 f_1 - \delta^2 f_0$$

and more generally:

$$\delta^{2r+1} f_{\frac{1}{2}} = \delta^{2r} f_1 - \delta^{2r} f_0 \text{ for } r = 1, 2, 3, \ldots$$

and substituting in (15) we have

$$f_p = f_0 + p(f_1 - f_0) + \frac{p(p-1)}{4}(\delta^2 f_0 + \delta^2 f_1) +$$
$$\frac{p(p-1)(p-\frac{1}{2})}{3!}(\delta^2 f_1 - \delta^2 f_0) + \ldots$$

then collecting like terms together, gives

$$f_p = (1 - p)f_0 + pf_1 + \left\{ \frac{p(p-1)}{4} - \frac{p(p-1)(p-\frac{1}{2})}{6} \right\} \delta^2 f_0 +$$

$$\left\{ \frac{p(p-1)}{4} + \frac{p(p-1)(p-\frac{1}{2})}{6} \right\} \delta^2 f_1 +$$

$$f_p = (1 - p)f_0 + pf_1 + \frac{p(p-1)}{24} \{6 - 4p + 2\} \delta^2 f_0 +$$

$$\frac{p(p-1)}{24} \{6 + 4p - 2\} \delta^2 f_1 + \ldots$$

Hence we have Everett's interpolation formula as

$$f_p = (1 - p)f_0 + pf_1 + \frac{p(p-1)(2-p)}{6} \delta^2 f_0 +$$

$$\frac{p(p-1)(p+1)}{6} \delta^2 f_1 + \ldots \qquad (16)$$

which can be seen, after the first two terms, to include only even order differences. It is possible, although the algebra becomes increasingly lengthy to obtain the coefficients of $\delta^4 f_0$, $\delta^4 f_1$, $\delta^6 f_0$, etc. as functions of p. Further coefficients are given in *I.A.T.*, page 56.

This formula, in general notation is usually given as

$$f_p = (1 - p)f_0 + pf_1 + E_2\delta_0^2 + F_2\delta_1^2 + E_4\delta_0^4 + F_4\delta_1^4 + \ldots$$

as on page 56 of *I.A.T.*

where the Everett coefficients $E_2, E_4, \ldots, F_2, F_4 \ldots$ are also given as functions of the interpolation coefficient 'p'.

1.10.1 As examples of Everett's formula in practice, we consider the two previous examples worked through using Bessel's formula in 1.9.1. and 1.9.3.

Worked example 1

Evaluate $f(0.36)$ given

	x	$f(x)$	$\delta^2 f$
x_0	0·3	0·878	34
→ 0·36			
x_1	0·4	0·782	34

using $x_p = x_0 + ph$

gives $p = 0.6$

Evaluating Everett's coefficients:

	Coefficients		
$1 - p$	$0 \cdot 4$	f_0	$0 \cdot 878$
p	$0 \cdot 6$	f_1	$0 \cdot 782$
$\dfrac{p(p - 1)(2 - p)}{6}$	(E_2) $-0 \cdot 056$	$\delta^2 f_0$	$0 \cdot 034$
$\dfrac{p(p - 1)(p + 1)}{6}$	$-0 \cdot 064$ (F_2)	$\delta^2 f_1$	$0 \cdot 034$

giving

$$f_p = (0 \cdot 4)(0 \cdot 878) + (0 \cdot 6)(0 \cdot 782) + (-0 \cdot 056)(0 \cdot 034) + (-0 \cdot 064)(0 \cdot 034)$$

with settings S.R.3; C.R.3; Acc. 6 gives

$$f(0 \cdot 36) = 0 \cdot 816\ 320 \text{ which is exact as before.}$$

1.10.2. Worked example 2 using the example in § 1.9.3 we have

$$
\begin{array}{cccc}
& x & f & \delta^2 f \\
x_0 & 0 \cdot 24 & 2 \cdot 972\ 71 & 67 \\
x_1 & 0 \cdot 25 & 2 \cdot 993\ 28 & 70
\end{array}
$$

and evaluate $f(0 \cdot 243)$

using $x_p = x_0 + ph$

$$\Rightarrow p = 0 \cdot 3$$

Obtaining the Everett coefficients:

	Coefficients		
$1 - p$	$0 \cdot 7$	f_0	$2 \cdot 972\ 71$
p	$0 \cdot 3$	f_1	$2 \cdot 993\ 28$
$\dfrac{p(p - 1)(2 - p)}{6}$	(E_2) $-0 \cdot 0595$	$\delta^2 f_0$	$0 \cdot 000\ 67$
$\dfrac{p(p - 1)(p + 1)}{6}$	$-0 \cdot 0455$ (F_2)	$\delta^2 f_1$	$0 \cdot 000\ 70$

giving

$$f(0.243) = (0.7)(2.972\ 71) + (0.3)(2.993\ 28) + (-\ 0.0595)(0.000\ 67) +$$
$$(-\ 0.0455)(0.000\ 70)$$

working with the settings

S.R. 5; C.R. 4; $\Rightarrow$ Acc. 9.

so that working to 9D we obtain

$f(0.243) = 2.978\ 809\ 285$ which when rounded-off

gives $f(0.243) = 2.978\ 81$ (5D) the same answer as previously.

1.10.3 Examples

In the following questions 1–5 use Bessel's second order interpolation formula.

1.

x	p	$f(x)$	$\delta^2 f$	$\delta^3 f$
2·2	0	18·168	192	
				6
2·3	1	21·037	198	

Find $f(x)$ to 3D when $x = 2.273$

2

x	p	$f(x)$	$\delta^2 f$	$\delta^3 f$
1·06	—	− 3·4791	13	
				1
1·07	—	− 3·3878	14	

Find the value of the function when $x = 1.063$, to 4D.

3.

x	p	$f(x)$	$\delta^2 f$	$\delta^3 f$
1·2	—	2·6354	− 659	
				14
1·3	—	2·2333	− 645	

Find $f(x)$ to 4D when $x = 1.235$

4.

x	p	$f(x)$	$\delta^2 f$	$\delta^3 f$
0·7	—	3·1519	146	
				19
0·8	—	3·2176	165	

Find $f(x)$ when $x = 0.767$ correct to 4D.

5.

x	p	$f(x)$	$\delta^2 f$	$\delta^3 f$
-0.13	—	$-0.588\ 99$	57	
				-3
-0.14	—	$-0.629\ 83$	54	

Find $f(x)$ to 5D when $x = -0.133$

6. (a) Use Bessel's third order interpolation polynomial to evaluate to 6D $\log_{10}(337.5)$ given

x	310	320	330	340	350	360
$\log_{10} x$	2·491 362	2·505 150	2·518 514	2·531 479	2·544 068	2·556 303

(b) Using Everett's formula, check the answer in (a); (Note the smaller amount of work and so the preference for Everett's formula under these circumstances).

1.11. CHOICE OF APPROPRIATE I.A.T. TABLES

In the examples considered in § 1.9 and § 1.10 we have followed a standard procedure and used either of the interpolation formulae as a check on the working performed with the other.

Considerable time was spent evaluating the Bessel or Everett coefficients for the appropriate values of the interpolating factor p. When the coefficients have been calculated it is possible to evaluate f_p as a continuous process (Σxy) on the machine. In practice we use the extensive tables giving the coefficients already evaluated for ranges of values of p.

The early tables in *I.A.T.* are examples of such tables. Much time will be saved by their use, but care and discretion is needed in deciding on the appropriate table. Regard must be paid to the accuracy required in the coefficient, which in turn is related to the magnitude of the difference with which it is associated.

Worked example

In order that we may consider the above problems, concerned with the use of *I.A.T.* tables, we consider the evaluation of sinh (1·860 55).

Taken from Chamber's Six Figure Mathematical Tables we have for the Hyperbolic Sine Function

x	$\sinh x$	δf	$\delta^2 f$	$\delta^3 f$
1·850	3·101 291			
		16 332		
1·855	3·117 623		77	
		16 409		2
1·860	3·134 032		79	
→ 1·860 55		16 488		− 1
1·865	3·150 520		78	
		16 566		2
1·870	3·167 086		80	
		16 646		
1·875	3·183 732			

(a) using $x_p = x_0 + ph$

$$\text{giving } p = \frac{x_p - x_0}{h}$$

$$p = \frac{1·860\ 55 - 1·860}{0·005}$$

$$\Rightarrow p = 0·11$$

(b) Using Everett's interpolation formula

$f_p = (1 - p)f_0 + pf_1 + E_2\delta^2 f_0 + F_2\delta^2 f_1$ (no further terms required in this case)

Then using $E_2 = \dfrac{-p(p - 1)(p - 2)}{3!}$ and $p = 0·11 \Rightarrow E_2 = -0·030\ 838\ 5$

and as $F_2 = \dfrac{(p + 1)p(p - 1)}{3!}$ and $p = 0·11 \Rightarrow F_2 = -0·018\ 111\ 5$

If we use these values of E_2, F_2 in the calculation, they would lead to terms involving 13 decimal places, which is far in excess of the accuracy that can be expected in the interpolate from functional values rounded off to 6D.

Here, we are faced with the basic problem in such calculations, namely, how many decimal places do we require in the interpolation coefficient. We have already suggested that the number of decimal places retained is related to the number of significant figures in the associated difference. But, before presenting a rule to assist in the selection of an appropriate number of decimal places in the interpolation coefficient it is instructive to work the example above, using various numbers of decimal figures in the coefficients. We will compare the results obtained.

1.11.1   First case: interpolation coefficients to 4D

From *I.A.T.* Table 6, p. 30–36, $p = 0.11$

the values given are $E_2 = -0.0309$; $F_2 = -0.0181$

$f_p = (0.89)(3.134\ 032) + (0.11)(3.150\ 520) + (-0.030\ 9)(0.000\ 079)$
$$(-0.0181)(0.000\ 078)$$
and working to 10D with S.R. 6, C.R. 4 $\Rightarrow$ Acc. 10.
$$f_p = 3.135\ 841\ 827\ 1$$
which rounded-off gives
$$\sinh (1.860\ 55) = 3.135\ 842\ (6D)$$

1.11.2   Second case: interpolation coefficients to 3D

$E_2 = -0.031$ and $F_2 = -0.018$
then
$$f_p = (0.89)(3.134\ 032) + (0.11)(3.150\ 520) + (-0.031)(0.000\ 079)$$
$$+ (-0.018)(0.000\ 078)$$
gives, working to 9D
$$f_p = 3.135\ 841\ 827$$
$$\Rightarrow \sinh (1.860\ 55) = 3.135\ 842 \quad (6D)$$

1.11.3   Third case: interpolation coefficients to 2D

Then $E_2 = -0.03$; $F_2 = -0.02$

thus
$$f_p = (0.89)(3.134\ 032) + (0.11)(3.150\ 520) + (-0.03)(0.000\ 079)$$
$$+ (-0.02)(0.000\ 079)$$
and working now to 8D gives
$$f_p = 3.135\ 841\ 75$$
$$\Rightarrow \sinh (1.860\ 55) = 3.135\ 842\ (6D)$$

1.11.4   Using Bessel's interpolation polynomial to evaluate $\sinh (1.860\ 55)$ will act as a check on the above calculations which used Everett's interpolation formula.

Bessel's formula: $f_p = f_0 + p\delta f_{\frac{1}{2}} + B_2(\delta^2 f_0 + \delta^2 f_1) + B_3\delta^3 f_{\frac{1}{2}} + \ldots$

The term $B_3\delta^3 f$ is omitted from the calculations. The difference $\delta^3 f_{\frac{1}{2}}$ has occurred through the rounding-off of the functional values and when associated with its interpolation coefficient the contribution it makes to the interpolate is negligible.

1.11.5   First case: using B_2 to 4D

From *I.A.T.* Table 6, page 31, we have
$$B_2 = - 0\cdot0245 \text{ (to 4D)}$$
then
$$f_p = 3\cdot134\,032 + (0\cdot11)\,(0\cdot016\,488) + (-0\cdot0245)\,(0\cdot000\,157)$$
with S.R. 6; C.R. 4 $\Rightarrow$ Acc. 10

gives $f_p = 3\cdot135\,841\,833\,5$

and $\sinh\,(1\cdot860\,55) = 3\cdot135\,842$ (6D)

(It is interesting to note that using Table 7 of *I.A.T.*, page 38, to obtain $B_3 = + 0\cdot0064$ and including the term $B_3\delta^3 f_{\frac{1}{2}}$ in the calculations gives $f_p = 3\cdot135\,841\,827\,1$ which is identical with the value in 1.11.1 and clearly demonstrates the equivalence of the two interpolation formulae.)

1.11.6   Second case: using B_2 to 3D

Then $B_2 = - 0\cdot025$ (3D)
$$f_p = 3\cdot134\,032 + (0\cdot11)\,(0\cdot016\,488) + (-0\cdot025)\,(0\cdot000\,157)$$
and working to 9D
$$\text{gives } f_p = 3\cdot135\,841\,755$$
and again
$$\sinh\,(1\cdot860\,55) = 3\cdot135\,842 \text{ (6D)}$$

So we see that all five cases considered give the same result to 6D. We now seek a rule which will always give sufficient accuracy and at the same time keep the calculations involved down to a minimum.

The above examples serve to show how the accuracy required in the result, together with the number of decimal places in the difference, will help decide the number of decimal places necessary in the interpolation coefficient.

1.11.7   Accuracy of interpolation coefficients

Butler and Kerr in *Introduction to Numerical Methods* suggest a simple rule to decide, within necessary accuracy, the number of decimal places in the interpolation coefficients. The rule is 'The number of decimal places in the interpolation coefficient should *not be less than* the number of significant figures in the associated (double) difference'. As a safety factor, it is worthwhile interpreting this rule as, '*one more than*' instead of 'not less than' and it was this interpretation that we used in the foregoing examples.

From the preamble of the *I.A.T.*, pages 14–23, specific suggestions are made with respect to the number of decimal places required in the interpolation coefficients. For example on page 17 is stated, that if Bessel's second-order interpolation formula is used and full accuracy is required, then if $(\delta^2 f_0 + \delta^2 f_1) < 500$ then Table 3 should be used giving B_2 to three decimal places. Similarly, exactly corresponding to the above rule, if $(\delta^2 f_0 + \delta^2 f_1) < 5000$, Table 6 is used to give B_2 to 4 D. Table 7, which gives B_2 up to 6D, as p varies in intervals of 0·001, is used for third and fourth degree interpolation.

For Everett's second-order interpolation, which as we saw earlier includes the effect of third differences, Table 6 is used if second differences are less than 5000.

We now summarize these and other conditions, as far as we are likely to need them. In the table below, the appropriate tables in *I.A.T.* are given, depending on the significant figures in the difference column.

1.11.8 Table for finding $B:E:F$ coefficients from *I.A.T.*

(a) Bessel's Interpolation Formula: $$\delta^2 f_0 + \delta^2 f_1 < 500$$	B_2 given to 3 D Table 3, *I.A.T.*
(b) Bessel: $$500 < \delta_0^2 + \delta_1^2 < 5000$$ Everett: $$\delta_0^2 < 5000,\ \delta_1^2 < 5000$$	B_2, E_2, F_2 all given to 4 D in Table 6, *I.A.T.*
(c) Bessel: $$\delta_2^2 + \delta_1^2 > 5000$$ Everett: $$\delta_1^2,\ \delta_0^2 > 5000$$	B_2 given to 6D, *I.A.T.* Table 7. E_2, F_2 given to 6D, *I.A.T.* Table 12.

This table can be used, to help decide the appropriate table in *I.A.T.* But we must consider the important problem of deciding which available formula to use and how many terms to retain. That is, we must consider the problem of *truncation errors*.

1.12 TRUNCATION ERRORS

These are errors which occur through the omission of terms from the inter-polation formula. By looking at these formulae, these errors may be reduced to negligible size, by retaining an adequate number of terms of the formula. Thus we are led to consider the term which will contribute less than $\frac{1}{2}$ unit to the final figure in the answer and truncate the formula immediately preceding the term satisfying such a condition.

Using the elementary methods of differential calculus it is not a difficult matter to evaluate the maximum value a difference (or sum of differences) may take in order that the contribution the associated term makes to the final result is less than $\frac{1}{2}$ unit in the final figure.

We work through two cases to demonstrate the method and then summarize the general results.

1.12.1 Case 1: using Bessel's interpolation formula:

$$f_p = f_0 + p\delta f_{\frac{1}{2}} + \frac{p(p-1)}{4}(\delta^2 f_0 + \delta^2 f_1) + \ldots$$

Suppose that we wish to find the maximum value that $(\delta^2 f_0 + \delta^2 f_1)$ may take, for value $0 \leqslant p \leqslant 1$ so that the contribution of the term is less than $\frac{1}{2}$ in the final figure.

i.e. $$\left|\frac{p(p-1)}{4}(\delta^2 f_0 + \delta^2 f_1)\right| < \frac{1}{2} \text{ for } 0 \leqslant p \leqslant 1$$

we thus require the maximum value of $|p(p-1)/4|$ so consider the function $C = p(p-1)$. For $0 < p < 1$ the function C is always negative and we obtain the maximum value of $|p(p-1)|$ if we obtain the minimum value of C.

Now $(\mathrm{d}C/\mathrm{d}p) = 2p - 1$ $(= 0$ for Max. or Min. values) and $(\mathrm{d}^2C/\mathrm{d}p^2) = 2$ which is > 0 so that there is here a minimum turning point.

i.e. Minimum value occurs when $p = \frac{1}{2}$ and $[p(p-1)/4] = -\frac{1}{16}$

$\therefore$ Maximum value of $\left|\dfrac{p(p-1)}{4}\right|$ is $\frac{1}{16}$

So the original condition becomes $\left|\frac{1}{16}(\delta_0^2 + \delta_1^2)\right| < \frac{1}{2}$

i.e. $$\delta_0^2 + \delta_1^2 < 8$$

giving $|\delta_0^2|$ and $|\delta_1^2| < 4$ for the term containing $(\delta_0^2 + \delta_1^2)$ to be omitted and linear interpolation will be permissible, as the truncation error will then, in general, be less than $\frac{1}{2}$ in the final place.

1.12.2 Case 2: using Everett's interpolation formula and the terms in $\delta^2 f$ from

$$f_p = (1-p)f_0 + pf_1 + E_2\delta^2 f_0 + F_2\delta^2 f_1 + \ldots$$

so that we require $|E_2\delta^2 f_0| + |F_2\delta^2 f_1| < \frac{1}{2}$

from *I.A.T.*, page 56, $$E_2 = \frac{-p(p-1)(p-2)}{3!},$$

i.e. we require the maximum value of $\left|\dfrac{-p(p-1)(p-2)}{3!}\right|$

$$\frac{\mathrm{d}(E_2)}{\mathrm{d}p} = k(3p^2 - 6p + 2)\,(= 0 \text{ for Max. or Min.})$$

giving $$p = \frac{3 \pm \sqrt{3}}{2} \text{ but for } 0 \leqslant p \leqslant 1 \text{ we take}$$

$$p = \frac{3 - \sqrt{3}}{2}$$

$$= 0{\cdot}4227$$

which using a machine, gives the minimum value of $E_2 = -0.0641$. (This value could have been found by direct reference to *I.A.T.* and inspecting Table 6, page 33 which shows E_2 with the above minimum value -0.0641.)

Similarly by inspection of *I.A.T.*, page 33, we find that F_2 has a minimum value of -0.0641.

$$\text{giving } |F_2| \text{ with maximum value } 0.0641$$

$\therefore$ If the second-order terms of Everett's formula are to be omitted

$$|E_2\delta^2 f_0| + |F_2\delta^2 f_1| < \tfrac{1}{2}$$

i.e.

$$|0.0641\ \delta^2 f_0| + |0.0641\ \delta^2 f_1| < \tfrac{1}{2}$$

giving

$$|\delta^2 f_0| + |\delta^2 f_1| < 8$$

or $\delta^2 f_0$ and $\delta^2 f_1 < 4$ in magnitude

Other limits of this kind may be calculated in a similar way. Below, these values have been tabulated, so that reference may be made, when we come to higher-order interpolation in the next part of the chapter.

1.12.3 The omission of terms needs care. Terms containing differences and higher-order differences may be omitted in any interpolation formula if their magnitudes are less than or equal to those given below.

	δ^2	δ^3	δ^4	δ^5	δ^6
Bessel's formula	4	60	20	500	100
Everett's formula	4	—	20	—	100
Gregory–Newton forward and backward formulae	4	8	12	16	21

$$\text{all for } 0 \leqslant p \leqslant 1$$

(In Bessel's formula where double differences occur, such as $(\delta^2 f_0 + \delta^2 f_1)$, the values in the above table should be doubled.)

In any table of differences, no matter what the magnitude of the function values, if the fourth-order differences were 18, then from the above table it would be permissible to omit the terms involving fourth-order differences and those of higher order in the Bessel and Everett interpolation formulae. Under these circumstances it would be quicker to use Everett's second-order formula, although Bessel's formula could then be used as a check.

1.12.4 Rounding-off errors need watching for build-up since we saw in Vol. I § 5.1.8 p. 124 that rounding off the function values leads to rounding-off errors in other parts of the table of differences. In fact, the errors build up as the order of differences increases and are greater than the error in the pivotal values, as demonstrated by the difference table at the end of § 5.1.8. Hence when using interpolation formulae the error in the final interpolate due to rounding-off errors may well be greater than the error in the function values.

The magnitude of the rounding-off error in the interpolate will also depend to some extent on the highest order of difference included in the interpolation polynomial.

In *I.A.T.*, page 15, are given the following maximum rounding-off errors for central difference interpolation formulae when the error in the function values $\leqslant \frac{1}{2}$ unit.

Highest order of difference	First	Third	Fifth
Maximum error	1·0	1·125	1·195

1.12.5   It is wise to keep truncation-error less than rounding-off error

When dealing with truncation error we saw how it is possible for us to control the magnitude of the truncation error by close consideration of the degrees of the interpolation polynomial. However, when rounding-off errors are involved we can in no way control or minimize the resulting final error thus produced in the interpolate f_p.

Hence, when both rounding-off errors and truncation errors are involved in a calculation it would seem advisable that the truncation error, which can be controlled, should be considerably less than the rounding-off error. In this connection, Hartree has suggested that it is advisable to retain contributions greater than 0·2 in the least significant digit, from the higher orders of differences.

Further, referring to the table in § 1.12.3, it is clear that after second-order difference formulae, the central difference formulae may be truncated at an earlier stage than the Gregory–Newton formulae. Then still considering the problem of keeping the truncation error much less than the rounding-off error, Butler and Kerr put it as follows: 'The Bessel and Everett interpolation coefficients decrease more rapidly than the Gregory–Newton coefficients and thus for truncation error to be less than the rounding-off error, Bessel and Everett formulae may be truncated at an earlier stage than the Gregory–Newton formula.' This is a further reason for preferring central-difference formulae wherever possible.

1.12.6   Stirling's interpolation formula

Before considering the problem of choice of formula for a particular calculation, we introduce another formula. It is Stirling's interpolation formula. This is not such an important practical formula as those already mentioned, but has important application when we consider differentiation and integration formulae, in later chapters.

Stirling's interpolation formula is:

$$f_p = f_0 + \tfrac{1}{2}p(\delta f_{-\frac{1}{2}} + \delta f_{\frac{1}{2}}) + \tfrac{1}{2}p^2\delta^2 f_0$$

$$+ \frac{p(p^2-1)}{2(3!)}(\delta^3 f_{-\frac{1}{2}} + \delta^3 f_{\frac{1}{2}}) + \frac{p^2(p^2-1)}{4!}\delta^4 f_0$$

$$+ \frac{p(p^2-1)(p^2-2^2)}{2(5!)}(\delta^5 f_{-\frac{1}{2}} + \delta^5 f_{\frac{1}{2}}) + \dots \tag{17}$$

or

$$f_p = f_0 + S_1(\delta f_{-\frac{1}{2}} + \delta f_{\frac{1}{2}}) + S_2\delta^2 f_0 + S_3(\delta^3 f_{-\frac{1}{2}} + \delta^3 f_{\frac{1}{2}}) + S_4\delta^4 f_0 + \dots$$

where the S_1, S_2, S_3, ..., are called Stirling interpolation coefficients and are given in *I.A.T.*, page 58.

It is to be noted that the differences used in the formula, lie on or immediately adjacent to the row with suffix x_0, etc.

x_{-1}

	$\delta f_{-\frac{1}{2}}$		$\delta^3 f_{-\frac{1}{2}}$		$\delta^5 f_{-\frac{1}{2}}$
x_0		$\delta^2 f_0$		$\delta^4 f_0$	
	$\delta f_{+\frac{1}{2}}$		$\delta^3 f_{\frac{1}{2}}$		$\delta^5 f_{\frac{1}{2}}$

x_1

1.12.7 Examples

Work through the examples in section 1.10.3 using Everett's interpolation formula.

1.13 CHOICE OF INTERPOLATION FORMULA

We have mentioned five interpolation formulae explicitly. You are always faced with the problem of which formula to use for any particular calculation.

We have already seen that when working from a table of differences in which there is a constant column of differences, we may conclude that the function values are exact values of a polynomial. Worked examples have verified that under these particular circumstances, if it is required, an exact value of the interpolate may be obtained. The results from § 1.9.1 and §1.10.1 serve to demonstrate the equivalence of the Bessel and Everett formula respectively, under these circumstances.

Also, in § 1.9 it was pointed out that the central-difference formulae involve the use of function values on either side of f_0, which are usually available. In § 1.12.3 it was explained, that by direct comparison both the Bessel and Everett coefficients are more rapidly convergent than those of the Gregory–Newton formulae. Then in § 1.12.5, when dealing with 'rounding-off' errors and truncation error, we again saw that the advantage lay with the central difference formulae. Taking into account all these considerations it is generally preferable to use a central-difference interpolation formula if possible.

In general there is little to choose between the amount of work involved in using either Bessel or Everett formulae. However, for third order interpolation and above, because odd-order differences are not involved in Everett's formula, neither do double differences (e.g. $\delta^2 f_0 + \delta^2 f_1$) have to be evaluated, it may be preferable to use Everett's formula. However, in practice almost certainly, Bessel's Formula would be used as a check.

1.14 MODIFIED DIFFERENCES OR 'THROWBACK'

This is a device or process, developed by L. J. Comrie, which saves time and labour. In its simplest form it is a simple modification of the Interpolation formulae, which gives a 'throwback' of fourth differences to second differences in the following way.

Consider Bessel's formula

$$f_p = f_0 + p\delta_{\frac{1}{2}} + B_2(\delta_0^2 + \delta_1^2) + B_3\delta_{\frac{1}{2}}^3 + B_4(\delta_0^4 + \delta_1^4) + \dots$$

in which, from earlier work or from *I.A.T.* p. 56, we have,

$$B_2 = \frac{p(p-1)}{4}; \qquad B_4 = \frac{\frac{1}{2}(p+1)\,p(p-1)\,(p-2)}{4!};$$

so that the combined contribution to the final interpolate, of the terms containing the second and fourth differences, is

$$\frac{p(p-1)}{4}\,(\delta_0^2 + \delta_1^2) + \frac{\frac{1}{2}(p+1)\,p(p-1)\,(p-2)}{24}\,(\delta_0^4 + \delta_1^4)$$

which can be written as

$$\frac{p(p-1)}{4}\left[(\delta_0^2 + \delta_1^2) + \frac{(p+1)\,(p-2)}{12}\,(\delta_0^4 + \delta_1^4)\right]$$

We now look more closely at the coefficient of the $(\delta_0^4 + \delta_1^4)$ term, if p lies within the range $0 < p < 1$. Under these conditions, $[(p+1)\,(p-2)/12]$ varies only from $-\frac{1}{6}$ to $-\frac{3}{16}$; taking the value $-\frac{3}{16}$ when $p = \frac{1}{2}$ and the value $-\frac{1}{6}$ when $p = 0$ or 1.

The maximum magnitude of this coefficient is $0\cdot1875$ when $p = \frac{1}{2}$. In particular the magnitude is greater than $0\cdot180$ for over half the range of p.

$$\left[\left|\frac{(p+1)\,(p-2)}{12}\right| = 0\cdot180 \text{ when } p = 0\cdot2 \text{ and } 0\cdot8 \text{ (see Fig. 1.14)}\right]$$

Consider the graph of the function $C(p) \equiv \dfrac{(p+1)\,(p-2)}{12}$.

The graph shows how small is the change in the value of the ordinate for values of p between 0 and 1. We require the most suitable single value of the function, to use as a coefficient, irrespective of the value of p, with the $(\delta^4 f_0 + \delta^4 f_1)$ term. This required value of the constant multiple can only be obtained following a more detailed analysis than would be appropriate here. Such analysis is contained in more advanced texts and proves that the best value of the constant is $-0 \cdot 184$. With this constant multiple of the fourth differences taken

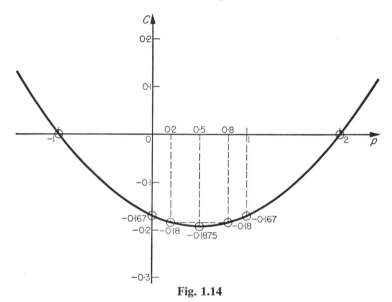

Fig. 1.14

from the second differences, the interpolation polynomial is truncated after the term containing the third order differences. There will, of course, be a slight contribution missing from the $\delta^4 f$ term, but using this value $-0 \cdot 184$ for the coefficient, it can be shown that, *provided the fourth difference is less than* 1000, the missing contribution is less than $\frac{1}{2}$ unit in the last significant figure of the calculation.

For the fuller explanations and proofs mentioned above the reader is referred to *Numerical Analysis* by Hartree; *Chambers' Six Figure Mathematical Tables Vol. II* by L. J. Comrie; and the *Introduction to I.A.T.*

Using these modified differences, Bessel's formula becomes

$$f_p = f_0 + p\delta f_{\frac{1}{2}} + B_2(\delta^2_m f_0 + \delta^2_m f_1) + B_3\delta^3 f_{\frac{1}{2}} \qquad (18)$$

where the notation is obvious and

$$\delta^2_m f_0 = \delta^2 f_0 - 0 \cdot 184\ \delta^4 f_0 \qquad (19)$$

In particular, the value of $0 \cdot 184\ \delta^4 f$ can be read off from Table 8, page 42, *I.A.T.* Similarly this 'throwback' device may be used with Everett's formula to give

$$f_p = (1 - p)f_0 + pf_1 + E_2\delta^2_m f_0 + F_2\delta^2_m f_1 \qquad (20)$$

The modified formulae discussed here are concerned with the 'throwback' of fourth differences to second differences. Similar modifications can be applied to higher-order differences, a list being given on page 57, *I.A.T.*

As an example on the use of modified differences, consider the following:

x	$f(x)$	δf	$\delta^2 f$	$\delta^3 f$	$\delta^4 f$
1·3	2·8561				
		22 064			
1·5	5·0625		10 832		
		32 896		3072	
1·7	8·3521		13 904		384
		46 800		3456	
1·9	13·0321		17 360		384
		64 160		3840	
2·1	19·4481		21 200		
		85 360			
2·3	27·9841				

(1·736 → 1·7 ... 1·9)

From this table evaluate $f(1·736)$ to 4D

using

$$x_p = x_0 + ph$$

$$p = \frac{x_p - x_0}{h}$$

$$= \frac{0·036}{0·2}$$

$$\Rightarrow p = 0·18$$

Now using Bessel's interpolation formula with modified differences

$$f_p = f_0 + p\delta f_{\frac{1}{2}} + B_2 (\delta_m^2 f_0 + \delta_m^2 f_1) + B_3 \delta^3 f_{\frac{1}{2}}$$

Then use Table 8, page 42, *I.A.T.*, to evaluate $0·184\ \delta^4$ and also use Table 7 for finding B_2, because, as stated earlier, the number of decimals in the interpolation coefficient should not be less than the number of significant figures in the associated difference,

using

$$\delta_m^2 f_0 = \delta^2 f_0 - 0·184\ \delta^4 f_0$$

$$= 1·3904 - 0·0071$$

$$= 1·3833$$

also

$$\delta_m^2 f_1 = 1·7360 - 0·0071$$

$$= 1·7289$$

giving

$$f_p = 8{\cdot}3521 + (0{\cdot}18)\,(4{\cdot}6800) + (-\,0{\cdot}036\,900)\,(1{\cdot}3833 + 1{\cdot}7289) +$$
$$(0{\cdot}007\,87)\,(0{\cdot}3456)$$
$$f_p = 9{\cdot}082\,379\,692$$
$$\therefore f(1{\cdot}736) = 9{\cdot}0824\;(4\text{D})$$

To compare the amount of computation involved we solve the same problem, without using modified differences, using Bessel's fourth-order interpolation formula

$$f_p = f_0 + p\delta f_{\frac{1}{2}} + B_2(\delta^2 f_0 + \delta^2 f_1) + B_3\delta^3 f_{\frac{1}{2}} + B_4(\delta^4 f_0 + \delta^4 f_1)$$

where B_2, B_3, B_4 are from Table 7.

then

$$f_p = 8{\cdot}3521 + (0{\cdot}18)\,(4{\cdot}6800) + (-\,0{\cdot}036\,900)\,(1{\cdot}3904 + 1{\cdot}7360)$$
$$+ (0{\cdot}007\,87)\,(0{\cdot}3456) + (0{\cdot}0066)\,(0{\cdot}0768)$$

giving

$$f_p = 9{\cdot}082\,362\,592$$
$$\Rightarrow f(1{\cdot}736) = 9{\cdot}0824\;(4\text{D})$$

1.15   SUMMARY

Section 1.1   Interpolation

1.2   The forward difference operator, Δ.

1.3   The shift operator, E.

1.4   The Gregory–Newton forward difference interpolation formula.

$$f_p = f_0 + \frac{p}{1!}\Delta f_0 + \frac{p(p-1)}{2!}\Delta^2 f_0 + \frac{p(p-1)\,(p-2)}{3!}\Delta^3 f_0 + \ldots$$

where 'p' is the interpolating factor and if 'h' is the interval of tabulation, then $x_p = x_0 + ph$.

1.5   The backward difference operato,r ∇.

1.6   The Gregory–Newton backward difference interpolation formula.

$$f_p = f_0 + p\nabla f_0 + \frac{p(p+1)}{2!}\nabla^2 f_0 + \frac{p(p+1)\,(p+2)}{3!}\nabla^3 f_0 + \ldots$$

1.8   Central difference operator, δ.

1.9   Bessel's interpolation formula.

$$f_p = f_0 + p\delta f_{\frac{1}{2}} + \frac{p(p-1)}{4}(\delta^2 f_0 + \delta^2 f_1) + \frac{p(p-1)\,p(p-\frac{1}{2})}{3!}\delta^3 f_{\frac{1}{2}}$$
$$+ \frac{(p+1)\,p(p-1)\,(p-2)}{2.4!}(\delta^4 f_0 + \delta^4 f_1) + \ldots$$

1.9.2   Interpolation and Allied Tables.

1.10   Everett's interpolation formula.

$$f_p = (1 - p)f_0 + pf_1 + \frac{p(p - 1)(2 - p)}{6}\,\delta^2 f_0 +$$
$$\frac{p(p - 1)(p + 1)}{6}\,\delta^2 f_1 + \ldots$$

1.11.8 Table showing appropriate *I.A.T.* for obtaining coefficients for Bessel and Everett formulae.

1.12.   Truncation errors.

1.12.3 Table showing the limits of the differences when truncating interpolation polynomials.

1.12.4 Rounding-off errors.

1.12.6 Stirling's interpolation formula.

$$f_p = f_0 + \tfrac{1}{2} p\,(\delta f_{-\frac{1}{2}} + \delta f_{\frac{1}{2}}) + \tfrac{1}{2} p^2 \delta^2 f_0$$
$$+ \frac{p(p^2 - 1)}{2.3!}\,(\delta^3 f_{-\frac{1}{2}} + \delta^3 f_{\frac{1}{2}}) + \frac{p^2(p^2 - 1)}{4!}\,\delta^4 f_0 + \ldots$$

1.13   Choice of interpolation formula in practice.

1.14   Modified differences or 'throwback'.

1.16   EXAMPLES

1. Using the following functional values for $x = 0.1(0.1)0.5$ find the value of the function for $x = 0.25$

x	0·1	0·2	0·3	0·4	0·5
$f(x)$	2·111	2·248	2·417	2·624	2·875

2. Using Bessel and Everett interpolation formulae, one to check the accuracy of the other, evaluate $f(0.543)$ given the following functional values:

x	0·1	0·2	0·3	0·4	0·5	0·6	0·7
$f(x)$	2·631	3·328	4·097	4·944	5·875	6·896	8·013

3. Evaluate, using an appropriate interpolation formula, $f(2.53)$ given the following table of values:

x	2·3	2·4	2·5	2·6	2·7	2·8
$f(x)$	16·371	17·788	19·309	20·940	22·687	24·556

4. Using five-figure tables tabulate sin x to 4D for $x = 0.4(0.2)2.0$, and obtain an approximate value of sin 1·234 from this table, using linear interpolation. Obtain a more accurate value for sin 1·234 from this table, using Everett's interpolation formula.

(Cambridge 1965).

5. Given the following values of sin x for $x = 0.15(0.02)0.23$

x	0·15	0·17	0·19	0·21	0·23
sin x	0·149 438	0·169 182	0·188 859	0·208 460	0·227 978

Evaluate sin x for $x = 0.197$

6. Using the methods described in sections 1.12.1 and 1.12.2 verify that $\delta^3 f_{\frac{1}{2}}$ must be less than 60 in Bessel's interpolation formula, for the contribution that this term makes to the final answer to be less than $\frac{1}{2}$ unit in the final figure.

7. Given the following values for sec x for $x = 1.250(0.005)1.270$ evaluate sec (1.263)

x	1·250	1·255	1·260	1·265	1·270
sec x	3·171 36	3·219 85	3·269 93	3·321 68	3·375 18

8. (i) Show that the value at $x = 2.76$ in the following table is 0·435 692.

(ii) Evaluate $f(x)$ at $x = 2.722\ 85$.

x	$f(x)$	$\delta^2 f$	$\delta^4 f$
2·5	0·490 358	2243	61
2·6	0·467 807	1956	46
2·7	0·447 212	1715	38
2·8	0·428 332	1512	29
2·9	0·410 964	1338	27
3·0	0·394 934	1191	21
3·1	0·380 095	1065	16

(A.E.B. 1959)

9. Using the following table of values for tanh x for $x = 2.20(0.01)2.26$ find the value of tanh (2.217) to 6D.

x	2·20	2·21	2·22	2·23
tanh x	0·975 743	0·976 218	0·976 683	0·977 140
x	2·24	2·25	2·26	
tanh x	0·977 587	0·978 026	0·978 457	

10. Tabulate $\log_{10}(\cosh x)$ for $x = 1\cdot5(0\cdot1)2\cdot0$, to 6D and form the difference table with differences up to the fourth order.

(i) Evaluate $\log_{10}(\cosh 1\cdot7326)$ by applying the Bessel (or a suitable alternative) interpolation formula.

(ii) Also, find $\cosh 1\cdot7326$ by linear interpolation from the tables and compare its common logarithm with the previous result. (A.E.B.)

11. Using fourth-degree interpolation formula on the following table:

x	1·7	1·9	2·1	2·3	2·5	2·7	2·9
$f(x)$	28·172	39·425	53·827	71·912	94·250	121·451	154·164

Evaluate $f(2\cdot13)$ to 3D.

12. From the following tabular values, using an appropriate interpolation polynomial, evaluate $f(0\cdot137)$

x	0·11	0·12	0·13	0·14	0·15	0·16
$f(x)$	1·135 53	1·150 53	1·166 00	1·181 94	1·198 38	1·215 30

13. Using the given table, compute $f(0\cdot2427)$.

x	$f(x)$
0·18	0·868 134
0·20	0·945 602
0·22	1·025 744
0·24	1·108 885
0·26	1·195 361
0·28	1·285 526
0·30	1·379 759
0·32	1·478 477

(A.E.B. 1965)

14. (a) Using fourth order Bessel formula and Table 11, page 43, *I.A.T.* for interpolation coefficient, evaluate $f(3\cdot45)$; where $f(x) \equiv \sinh x$.

x	3·1	3·2	3·3	3·4	3·5
$f(x)$	11·076 45	12·245 88	13·537 88	14·965 36	16·542 63
x	3·6	3·7	3·8		
$f(x)$	18·285 46	20·211 29	22·339 41		

(b) Check the answer in part (a) by using Everett's formula with modified differences.

15. The table gives values of e^x and corresponding differences. Using the table calculate the value of e^x when $x = 3\cdot2382$. Check your solution using Table Vc in *Chambers' Shorter Six-Figure Tables*.

x	e^x	δ^2	δ^4	δ^6
2·8	16·4446	6 600	265	9
3·0	20·0855	8 061	324	11
3·2	24·5325	9 846	394	21
3·4	29·9641	12 025	485	11
3·6	36·5982	14 689	587	31
3·8	44·7012	17 940	720	30
4·0	54·5982	21 911	833	23

(A.E.B.)

16.

x	$f(x)$	δ^2	δ^4	δ^6	δ^8
0·5	17·636	−25	563	−10	3
0·6	117·636	100	542	+10	4

If $f(x)$ and all its significant even order differences are given in the above table, for $x = 0\cdot5$ and $0\cdot6$. Using (a) Everett's interpolation formula and (b) any other interpolation formula obtain $f(0\cdot543\ 21)$.   (Cambridge).

2
Inverse Interpolation

2.1 INTRODUCTION

In Chapter 1, given values of the independent variable x, at equal intervals, together with the associated function values $f(x)$, and using interpolation polynomials, we were able to find values of the function for values of the independent variable lying between the tabulated values. In this chapter we consider the inverse problem, which is, given a non-tabulated value of the function to find the associated value of the independent variable. Such a process is called *inverse interpolation*.

2.2 INVERSE LINEAR INTERPOLATION

This is not a new process for when using trigonometrical tables, in particular, natural sines, cosines and tangents, if given the value of a ratio and the mean differences, we can find the value of the angle, i.e. the process of inverse interpolation has been performed.

Worked example

e.g. Consider $\tan \theta = 1.258\,67$ to find θ.
 The nearest entry is $\theta = 51° 30'$ given by $\tan \theta = 1.257\,17$ then using the mean differences we have

$$\tan 51° 32' = 1.258\,67$$

$$\text{i.e. } \theta = 51° 32'$$

We have used the fact that $f_p = f_0 + p\delta f_m$ (where δf_m is a mean difference). The values of f_p, f_0 and δf_m determine 'p' uniquely. Then the value of θ is determined using

$$\theta_p = \theta_0 + p.h$$

2.2.1 Consider another example given:

x	$f(x)$	δf	$\delta^2 f$
0·73	1·5329		
		147	
0·74	1·5476		2
?	[→ 1·5537]	149	
0·75	1·5625		2
		151	
0·76	1·5776		

Find the value of x for which $f(x) = 1\cdot5537$.

Method A

As second-order differences are sufficiently small we may use the linear interpolation formula.

$$f_p = f_0 + p \cdot \delta f_{\frac{1}{2}}$$

then as f_p, f_0, and $\delta f_{\frac{1}{2}}$ are known we can find 'p'

for
$$p \cdot \delta f_{\frac{1}{2}} = f_p - f_0$$

$$p(0\cdot0149) = 1\cdot5537 - 1\cdot5476$$

$$\text{gives } p = \frac{0\cdot0061}{0\cdot0149}$$

$$\Rightarrow p = 0\cdot409\ 396$$

and already we have a very important problem to consider. In inverse interpolation, how accurate can we expect 'p' to be? We shall examine the problem more fully in § 2.3 below.

However, in the present problem the function values are given to 4D so we will assume it reasonable to obtain x_p to four decimal places, though whether this is a justifiable attempt cannot be considered until later.

Here the values of the independent variable x are given to two decimal places and the interval h is 0·01. So in using $x_p = x_0 + ph$ if 'p' is rounded to 3D the ph term will then involve five decimal places and x_p can then be rounded to 4D.

$$\text{So } x_p = x_0 + ph \text{ gives}$$

$$x_p = 0\cdot74 + (0\cdot409)(0\cdot01)$$

$$x_p = 0\cdot744\ 09$$

$$\Rightarrow x_p = 0\cdot7441\ (4D)$$

Method B

Another method of doing the above example is instructive for it involves finding 'p' by the 'build-up' method on a machine and obtaining the required function value in the Accumulator. This method also incorporates one simple automatic check in the calculations.

Method on machine:

(i) Move carriage to the right and clear all registers.

(ii) Set f_0 in the Acc. and clear S.R.

(iii) Set $\delta f_{\frac{1}{2}}$ in S.R.

(iv) Add in $\delta f_{\frac{1}{2}}$ and verify that it gives f_1 (the check).

(v) Subtract the $\delta f_{\frac{1}{2}}$ and move the carriage one place to the left, and clear the C.R.

(vi) 'Build-up' the required value of f_p in the Acc.

(vii) The corresponding value of 'p' is then in the C.R.

	S.R.	C.R.	Acc.
Clear all registers and set carriage to the right. Set 1·5476 in the Acc.	1·5476	1·000 000 0	1·547 600 000 00
Add in $\delta f_{\frac{1}{2}}$ (the check f_1)	0·0149	2·000 000 0	1·562 500 000 00
Subtract $\delta f_{\frac{1}{2}}$ and move carriage one place to the left. Clear C.R.	0·0149	0·000 000 0	1·547 600 000 00
'Build-up' the required value of f_p in the Acc. Read off 'p' in C.R.	0·0149	'p' 0·409 395 9	1·553 699 998 91

again thus $x_p = 0\cdot7441$ (4D) if 'p' is rounded to 2D.

2.3   ACCURACY OF THE CALCULATED INTERPOLATING FACTOR p.

2.3.1 The possible error in x resulting from an error in the function value is first considered. As we have seen, if the function value is rounded off to n decimal places, the error lies in the range $\pm \frac{1}{2} \times 10^{-n}$.

Consider this graphically from fig. 2.3.1.

where $\tan \theta = f'(x)$, we have the approximation

$$\delta x = \frac{\frac{1}{2} \times 10^{-n}}{f'(x)}$$

and more generally since the error in $f(x)$ may lie in the range $\pm \frac{1}{2} \times 10^{-n}$ we have

$$\delta x \text{ (the error in } x) = \frac{\pm \frac{1}{2} \times 10^{-n}}{f'(x)} \tag{1}$$

Note that if $|f'(x)| > 1$ then the error in x is less than the error in the function and correspondingly if $|f'(x)| < 1$ then the error in x will be greater than the error in $f(x)$.

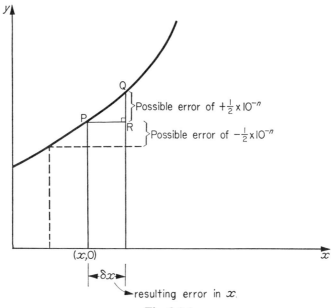

Fig. 2.3.1

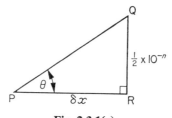

Fig. 2.3.1(a).

However, the difficulty of using this result, is that the function will rarely be known explicitly and so $f'(x)$ cannot be evaluated.

2.3.2 The related accuracy of the interpolating factor p, is developed by making use of the above result in (1).

Now from Taylor's Theorem

$$f(x + h) = f(x) + hf'(x) + \tfrac{1}{2} h^2 f''(x) + \ldots$$

Neglecting terms involving h^2 and higher powers of h

$$f'(x) \simeq \frac{1}{h} \{f(x + h) - f(x)\}*$$

$$\text{i.e. } f'(x) \simeq \frac{1}{h} \delta f_{\frac{1}{2}} \tag{2}$$

and on substituting (2) in (1) gives

$$\delta x \text{(the error in } x) = \frac{\pm \tfrac{1}{2} \times 10^{-n}}{\delta f_{\frac{1}{2}}} \cdot h \tag{3}$$

To obtain the related error in p, we use the method of 'small errors' from the differential calculus.

$$\frac{dx}{dp} \simeq \frac{\delta x}{\delta p}$$

$$\Rightarrow \delta p \text{ (i.e. the error in } p) \simeq \delta x \cdot \frac{dp}{dx} \tag{4}$$

But as $\qquad x_p = x_0 + ph$ we have $\dfrac{dx}{dp} = h$

and so $\qquad \delta p \simeq \delta x \cdot \dfrac{1}{h}$

giving $\qquad \delta p \simeq \dfrac{\pm \tfrac{1}{2} \times 10^{-n}}{\delta f_{\frac{1}{2}}} \tag{5}$

Then using (5) for each example involving inverse interpolation we may estimate quite simply, the error in p due to the rounding-off errors of the function values, by dividing the maximum possible error in the function by the first difference.

We now return to the example in 2.2 and apply (5) to give

$$\text{error in } p \simeq \frac{\pm \tfrac{1}{2} \times 10^{-4}}{0 \cdot 0149}$$

$$\simeq \frac{\pm \tfrac{1}{2} \times 10^{-4}}{1 \cdot 49 \times 10^{-2}}$$

$$\simeq \pm \tfrac{1}{3} \times 10^{-2}$$

* A fuller discussion on this and other approximations for $f'(x)$ is given later in Chapter 5.

$$\text{i.e. error} \simeq \pm\, 0\cdot0033$$
$$\therefore \quad p = 0\cdot4094 \pm 0\cdot0033$$
$$\text{i.e. } 0\cdot4127 > p > 0\cdot4061$$

showing that p may be correctly rounded to $0\cdot41$ (2D).

Thus x_p may be accurately given as $0\cdot7441$ (4D). Further it verifies that our arbitrary decision to attempt to compute x_p to 4D is attainable in this case.

It is important that the reader should, in future, use (5) and evaluate the maximum possible error in p before becoming involved in computations of p to more decimal places than are warrantable.

In the particular example above, the function of x we used was $f(x) \equiv x^2 + 1$. We may check the answer by substituting for $x_p = 0\cdot7441$.

2.4 SECOND-ORDER INVERSE INTERPOLATION

To demonstrate the result (5) and at the same time introduce higher order inverse interpolation consider the following example.

x	$f(x)$	δf	$\delta^2 f$	$\delta^3 f$
1·73	9·080 90			
		20 891		
1·74	9·289 81		364	
		21 255		3
1·75	9·502 36		367	
$\longrightarrow$ 9·620 09	21 622		4	
1·76	9·718 58		371	
		21 993		6
1·77	9·938 51		377	
		22 370		
1·78	10·162 21			

to find x when $f(x) = 9\cdot620\ 09$

(a) Using (5) we first consider the accuracy of p

$$\text{error in } p = \frac{\pm \tfrac{1}{2} \times 10^{-5}}{0\cdot216\ 22}$$
$$= \pm \tfrac{1}{4} \times 10^{-4} \text{ or } \pm 0\cdot000\ 025$$

and we will retain 5 decimal places in p.

(b) We consider two methods of finding the value of x. The first, method A, is an iterative or loop process and is the method usually preferred in practice.

In this method successive approximations to 'p' become more and more accurate. Also as third-order differences are negligible we use Bessel's second-order interpolation formula.

2.4.1 Method A

(i) from the above we retain five decimal places in calculating p,

(ii) as $500 < \delta^2 f_0 + \delta^2 f_1 < 5000$ from § 1.11.8 we use Table 6, *I.A.T.*, p. 30 *et seq*,

(iii) the first approximation to the interpolating factor, with an obvious notation will be $p^{(1)}$ and is found by linear interpolation.

$$\text{i.e. } f_p = f_0 + p \cdot \delta f_{\frac{1}{2}}$$
$$9{\cdot}620\ 09 = 9{\cdot}502\ 36 + p^{(1)} \cdot (0{\cdot}216\ 22)$$
$$\Rightarrow p^{(1)} = 0{\cdot}544\ 49$$

(iv) then using $p^{(1)} = 0{\cdot}544\ 49$ from Table 6 the first Bessel interpolation coefficient $B_2^{(1)} = -0{\cdot}0620$,

(v) the second approximation of the interpolating factor, $p^{(2)}$ is found using the second-order formula and the value $B_2^{(1)}$ above:

$$f_p = f_0 + p^{(2)} \cdot \delta f_{\frac{1}{2}} + B_2^{(1)}(\delta^2 f_0 + \delta^2 f_1) \text{ gives}$$
$$9{\cdot}620\ 09 = 9{\cdot}502\ 36 + p^{(2)}(0{\cdot}216\ 22) + (-0{\cdot}0620)(0{\cdot}007\ 38)$$

and using the machine we obtain

$$p^{(2)} = 0{\cdot}546\ 61$$

(vi) again using Table 6, *I.A.T.* with $p^{(2)} = 0{\cdot}5466$ gives $B_2^{(2)} = -0{\cdot}06200$ and the loop is complete because the value $B_2^{(2)}$ is the same as $B_2^{(1)}$ and further working only repeats identical working.

$$\therefore p = 0{\cdot}5466 \ (4D)$$

(vii) Then using $x_p = x_0 + ph$

$$x_p = 1{\cdot}75 + (0{\cdot}5466)(0{\cdot}01)$$
$$x_p = 1{\cdot}755\ 466$$

2.4.2 Full accuracy of the interpolating factor p

It is interesting to use the methods developed in Chapter 1 to find the maximum accuracy of the above answers. For we have, using the result from 2.4.1, that

$$p = 0{\cdot}546\ 61 \pm 0{\cdot}000\ 025$$

and so p lies in the range $0{\cdot}546\ 635$ and $0{\cdot}546\ 585$ and

$$\text{i.e. } 0{\cdot}546\ 635 > p > 0{\cdot}546\ 585$$

and can be rounded off correctly as

$$0·5466 \text{ (4D)}$$

Further

$$\text{as } x_p = 1·75 + (0·01)(0·546\ 61 \pm 0·000\ 025)$$

$$= 1·755\ 466\ 1 \pm 0·000\ 000\ 25$$

$$= 1·755\ 466 \text{ correct to 6D}$$

2.4.3 Method B

Again we use Bessel's interpolation formula truncated after the term containing the second-order differences

$$f_p = f_0 + p \cdot \delta f_{\frac{1}{2}} + B_2(\delta^2 f_0 + \delta^2 f_1)$$

in which f_p, f_0, $\delta f_{\frac{1}{2}}$, $\delta^2 f_0$ and $\delta^2 f_1$ are all known.

Then substituting $\dfrac{p(p-1)}{4}$ for B_2 gives a quadratic in p.

i.e. $9·620\ 09 = 9·502\ 36 + p(0·216\ 22) + \dfrac{p(p-1)}{4}(0·007\ 38)$

$$0·117\ 73 = p(0·216\ 22) + p(p-1)(0·001\ 845)$$

$$\therefore\ 0·001\ 845 p^2 + 0·214\ 375 p - 0·117\ 73 = 0$$

giving $\qquad 1845 p^2 + 214\ 375 p - 117\ 730 = 0$

$$369 p^2 + 42\ 875 p - 23\ 546 = 0$$

$$\therefore p = \frac{-42\ 875 \pm \sqrt{[(42\ 875)^2 + 4(369)(23\ 546)]}}{738}$$

where the accuracy of this calculation will depend on the capacity of the machine being used.

$$p = \frac{-42\ 875 \pm 43\ 278·39}{738}$$

giving $p = 0·546\ 599$ taking the positive root because $0 < p < 1$
Thus for consideration of full accuracy we have

$$p = 0·546\ 599 \pm 0·000\ 025$$

$$0·546\ 574 < p < 0·546\ 624$$

$$\text{giving } p = 0·5466 \text{ (to 4D)}$$

In general, this method takes longer than the iterative process of Method A and consequently is not often used in practice.

2.5 EXAMPLES ON SECOND-ORDER INVERSE INTERPOLATION

1. Use the following table to find x when $f(x) = 2 \cdot 3043$

x	$f(x)$	$\delta^2 f$
1·36	2·2947	1
1·37	2·3138	2

2. From the following table of function values:

x	1·31	1·32	1·33	1·34	1·35
$f(x)$	1·006 74	1·032 24	1·058 02	1·084 06	1·110 38

find x such that $f(x) = 1 \cdot 065\ 02$,

(i) by the method of obtaining a quadratic in p and solving it, and then finding x_p. What particular precaution is required when finding p by this method in this example?

(ii) Check the answer from (i) using the iterative process.

3. Using the following tabulated values and inverse interpolation

x	$f(x)$	$\delta^2 f$
1·051	7·377 94	163
1·061	7·456 69	165

find the value of x for which the function value is 7·425 18.

4. From the following tabular values of sinh x

x	1·850	1·855	1·860	1·865	1·870
sinh x	3·101 291	3·117 623	3·134 032	3·150 520	3·167 086

compute the value of x such that sinh $x = 3 \cdot 143\ 915$

5. Use the following table to calculate x when $f(x) = 3 \cdot 1942$

x	p	$f(x)$	$\delta^2 f$
0·7	0	3·1519	146
0·8	1	3·2176	165

6. From the following tabular values, form the difference table and then find x when $f(x) = 3 \cdot 155\ 09$.

x	0·53	0·55	0·57	0·59	0·61
$f(x)$	3·148 77	3·151 63	3·157 48	3·166 39	3·178 39

(N.B. You will find in this example, that $(\delta^2 f_0 + \delta^2 f_1)$ is greater than 500 but less than 5000 and so from the table in § 1.11.8 we use Table 6, *I.A.T.*, page 30, for finding B_2).

7. Tabulate the $f(x) \equiv 1\cdot7x^3 - 3\cdot6x^2 + 0\cdot7x + 6$ for $x = 1\cdot53(0\cdot01)1\cdot57$. Then using this table and inverse interpolation find x, when $f(x) = 4\cdot779\,348$.

8. Evaluate and tabulate $f(x) \equiv 1\cdot2x^3 - 3x^2 + 2\cdot7x + 4$ for $x = -1\cdot57\,(-0\cdot03) - 1\cdot69$ working to 5D. Then find x such that $f(x) = -13\cdot656\,92$.

2.6   THIRD-ORDER INVERSE INTERPOLATION

In the examples considered so far we have needed only to use Bessel's second-order formula. We now consider examples in which third-order differences must be taken into account. We shall then be able to use Everett's formula in this iterative process.

Worked example

The following function values have been rounded off to 2D.

x	$f(x)$	δf	$\delta^2 f$	$\delta^3 f$	$\delta^4 f$
1·7	28·17				
		1126			
1·9	39·43		314		
		1440		54	
2·1	53·83		368		4
$\longrightarrow$	56·89	1808		58	
2·3	71·91		426		2
		2234		60	
2·5	94·25		486		
		2720			
2·7	121·45				

We find the value of x for which $f(x) = 56\cdot89$.

From the previous chapter we know that Everett's second-degree formula automatically covers the third-order differences, whereas if we used Bessel's formula we must include the term $B_3 \delta^3 f_{\frac{1}{2}}$.

Proceeding with the method exactly as in previous examples.

For accuracy of p

(i)   error in $p = \dfrac{\frac{1}{2} \times 10^{-2}}{18 \cdot 08}$

$\simeq \dfrac{\frac{1}{2} \times 10^{-2}}{2 \times 10}$

$\simeq \frac{1}{4} \times 10^{-3}$ so we retain 4 decimal places in p.

(ii) Since both $\delta^2 f_0$ and $\delta^2 f_1 < 5000$ we use Table 6, *I.A.T.* to obtain values of E_2, F_2.

(iii) Then using Everett's formula:

$$f_p = f_0 + p \cdot \delta f_{\frac{1}{2}} + E_2 \cdot \delta^2 f_0 + F_2 \delta^2 f_1.$$

We find $p^{(1)}$ as before, from the linear interpolation formula.

$$f_p = f_0 + p^{(1)} \cdot \delta f_{\frac{1}{2}}$$

$$56 \cdot 89 = 53 \cdot 83 + p^{(1)} \cdot 18 \cdot 08$$

$$p^{(1)} = 0 \cdot 1692 \text{ and using Table 6, } I.A.T.$$

$$E_2^{(1)} = - 0 \cdot 0428 \text{ and } F_2^{(1)} = - 0 \cdot 0274$$

giving $p^{(2)}$ from

$$56 \cdot 89 = 53 \cdot 83 + p^{(2)} \cdot 18 \cdot 08 + (- 0 \cdot 0428)(3 \cdot 68) + (- 0 \cdot 0274)(4 \cdot 26)$$

$$\Rightarrow p^{(2)} = 0 \cdot 1844$$

giving $E_2^{(2)} = - 0 \cdot 0455$ and $F_2^{(2)} = - 0 \cdot 0297$

Then $p^{(3)}$ is found from

$$56 \cdot 89 = 53 \cdot 83 + p^{(3)} \cdot 18 \cdot 08 + (- 0 \cdot 0455)(3 \cdot 68) + (- 0 \cdot 0297)(4 \cdot 26)$$

$$\Rightarrow p^{(3)} = 0 \cdot 1855$$

giving $E_2^{(3)} = - 0 \cdot 0457$ $F_2^{(3)} = - 0 \cdot 0299$

Then

$$56 \cdot 89 = 53 \cdot 83 + p^{(4)} \cdot 18 \cdot 08 + (- 0 \cdot 0457)(3 \cdot 68) + (- 0 \cdot 0299)(4 \cdot 26)$$

giving $p^{(4)} = 0 \cdot 18560$ and the values of E_2 and F_2 remain unchanged and the loop is complete. The full accuracy of p is that it will lie in the range

$$0 \cdot 185\ 60 \pm 0 \cdot 000\ 25$$

between $0 \cdot 185\ 85$ and $0 \cdot 185\ 35$.

C

For value of x_p

Using $x_p = x_0 + ph$

Thus x_p must lie between $2 \cdot 1 + (0 \cdot 2)(0 \cdot 185\ 85)$ and $2 \cdot 1 + (0 \cdot 2)(0 \cdot 18\ 535)$

$\therefore 2 \cdot 137\ 070 < x_p < 2 \cdot 137\ 70$

so that greatest accuracy permissible gives $x_p = 2 \cdot 137$ which you can check by considering the question and answer to No. 12, § 1.15 Chapter 1, or by direct interpolation from the given table.

2.6.1 Subtabulation near the value of p simplifies working. It is intuitive from the theory of difference tables that if third-order differences are large then the lower-order differences will be relatively larger. Consequently, as demonstrated, we can expect more steps in the iterative process than is the case when third-order differences are negligible. However, if even higher-order interpolation is necessary the method so far demonstrated is always applicable. In practice though, in most cases where higher-order formulae are involved, the working can be much simplified if subtabulation is performed near to the value of p given by the approximation $f_p = f_0 + p\delta f_{\frac{1}{2}} + B_2(\delta^2 f_0 + \delta^2 f_1)$.

By subtabulation we mean a series of direct interpolations carried out between two adjacent function values of the main table, using necessarily a smaller tabular interval, usually $\frac{1}{2}$, $\frac{1}{5}$ or $\frac{1}{10}$ of the given interval h. This set of new computed interpolates should then be checked by differencing. For a fuller discussion of subtabulation the reader is referred to the booklet *Sub-tabulation* (H.M.S.O.) and *Numerical Analysis* by D. R. Hartree.

2.7 SOLUTION OF EQUATIONS

One very important application of inverse interpolation is solving equations. The equations may be purely algebraic or involve other functions such as circular and hyperbolic functions. In Vol I, Chapters 4 and 7 we have already given some methods of solution of these equations. Inverse interpolation provides an additional method which in some cases may be better than those previously studied. For example in the case of a function which is easy to tabulate but for which the computation of the derivative is cumbersome, inverse interpolation may be better than Newton's method of approximation to the roots of an equation.

Worked example

Consider the equation $x \sin x = 1$. The function $f(x) \equiv x \sin x - 1$ was given and drawn in Vol 1, Chapter 3, § 3.6. Then we saw that there were an infinite number of roots. We now find the smallest positive root, which we know is approximately $x = 1 \cdot 1$. Therefore we must tabulate $f(x) \equiv x \sin x - 1$

around $x = 1\cdot1$. Using *Chamber's Six Figure Tables* and rounding-off the function values to 6D, we have:

x	$\sin x$	$f(x) \equiv x \sin x - 1$	δf	$\delta^2 f$	$\delta^3 f$	$\delta^4 f$
1·06	0·872 355	− 0·075 304				
			27 819			
1·08	0·881 958	− 0·047 485		− 6		
			27 813		− 23	
1·10	0·891 207	− 0·019 672		− 29		− 2
$f(x) = 0$ ⟶			27 784		− 25	
1·12	0·900 100	+ 0·008 112		− 54		− 2
			27 730		− 27	
1·14	0·908 633	0·035 842		− 81		+ 3
			27 649		− 24	
1·16	0·916 803	0·063 491		− 105		
			27 544			
1·18	0·924 606	0·091 035				

By inspecting the table we see that a root of $f(x) = 0$ lies between $x = 1\cdot10$ and $x = 1\cdot12$. So we find x for $f(x) = 0$ by inverse interpolation. In this way the problem which began as requiring the solution of an equation is transformed into an inverse interpolation problem.

For accuracy of p

(i) Error in p using $\dfrac{\frac{1}{2} \times 10^{-6}}{\delta f_{\frac{1}{2}}}$

$$= \frac{\frac{1}{2} \times 10^{-6}}{0\cdot03}$$

$$= \frac{1}{6} \times 10^{-4}$$

and we will retain **6** figures in the working

(ii) Using Table 6, *I.A.T.*, for E_2 and F_2.

(iii) First approximation of $p^{(1)}$ using linear interpolation $f_p = f_0 + p \cdot \delta f_{\frac{1}{2}}$,

$f_p = 0, f_0 = - 0\cdot019 \ 672, \delta f_{\frac{1}{2}} = 0\cdot027 \ 784.$

$\therefore 0 = - 0\cdot019 \ 672 + p^{(1)} (0\cdot027 \ 784)$

$\Rightarrow p^{(1)} = 0\cdot708 \ 033$

(iv) Second approximation of $p^{(2)}$ using Everett's formula

from Table 6, $E_2^{(1)} = -0.0445$ and $F_2^{(1)} = -0.0589$

giving $0 = -0.019\ 672 + p^{(2)}(0.027\ 784) + (-0.0445)\ (0.000\ 029)$

$$+\ (-0.058\ 9)\ (0.000\ 054)$$

$\Rightarrow p^{(2)} = 0.707\ 872$

and from Table 6 we see that E_2 and F_2 are unchanged and so the loop is complete.

Thus $p = 0.707\ 872 \pm 0.000\ 016$

and p lies in the range

$$0.707\ 888 > p > 0.707\ 856$$

i.e. p may be accurately rounded to 0.7079 (4D).

For value of x_p

$$x_p = x_0 + ph \text{ giving } x_p = 1.114\ 158$$

$$\therefore x_p = 1.114\ 16 \text{ radians (5D)}$$

2.7.1 **Worked example.** An accuracy check of the above result, is by substitution in the given equation. The value of $\sin x$ will be found by direct interpolation from Table IIIE page 196 *Chamber's Six Figure Tables.*

x	$\sin x$	δf	$\delta^2 f$
1·113	0·897 029		
		441	
1·114	0·897 470		0
		441	
1·115	0·897 911		1
		440	
1·116	0·898 351		

As $\delta^2 f < 4$ linear interpolation is permissible.

$x_p = 1.114\ 157$, $x_0 = 1.114$, $h = 0.001$, $\Rightarrow p = 0.157$

$\therefore f_p = 0.897\ 470 + (0.157)\ (0.000\ 441)$

giving $\sin (1.114\ 157) = 0.897\ 539$ (6D)

and using this value for $\sin x$ in $f(x) \equiv x \sin x - 1$ gives a residual of $-0.000\ 000\ 6$, demonstrating a very satisfactory degree of accuracy.

2.8   MISCELLANEOUS EXAMPLES

1. A function $f(x)$ of x is tabulated at equal intervals so that the tabulated values of $f(x)$ bridge a constant A. Describe briefly one direct and one indirect method of solving $f(x) = A$.

Illustrate one of the methods by solving $f(x) = 3\cdot0000$ where $f(x)$ is given by the following table:

x	13·92	13·94	13·96	13·98	14·00	14·02
$f(x)$	0·4004	1·4186	2·4573	3·5170	4·5982	5·7011

(Cambridge)

2. Using the following table, where $f(x)$ is correct to 6D,

x	1·63	1·65	1·67	1·69	1·71
$f(x)$	11·924 583	12·509 210	13·114 741	13·741 692	14·390 577

(i) evaluate the accuracy of the interpolating factor.

(ii) find x, as accurately as possible, where
$$f(x) = 12\cdot848\ 733.$$

3. Tabulate $f(x) \equiv 4\cdot766x^3 + 9\cdot543x^2 + 6\cdot127x + 37\cdot354$ for $x = -3\ (0\cdot1) - 2\cdot3$.

Then use the table to find x when $f(x) = 0$ by the method of inverse interpolation.

4. From Vol. I, § 3.6.1 we have for $x^4 + 3x - 1 = 0$ that one of the roots lies near $x = -1\cdot54$. By tabulating the function $x^4 + 3x - 1$ for $x = -1\cdot57\ (0\cdot01) - 1\cdot51$ and using inverse interpolation find the root near $-1\cdot54$ to 5D.

5. The table gives values of e^x and corresponding differences. Using the table calculate the value of e^x when $x = 3\cdot2382$. Check your solution using Table Vc in *Chamber's Shorter Six-figure Tables*.

x	e^x	δ^2	δ^4	δ^6
2·8	16·4446	6 600	265	9
3·0	20·0855	8 061	324	11
3·2	24·5325	9 846	394	21
3·4	29·9641	12 025	485	11
3·6	36·5982	14 689	587	31
3·8	44·7012	17 940	720	30
4·0	54·5982	21 911	883	23

By the method of inverse interpolation, evaluate x when $e^x = 30$, using the values in the table above.

(A.E.B.)

6. Show graphically that the equation $15x - 10 \sinh x = 1$ has a positive root and evaluate it correct to 6S.

(A.E.B.)

7. Show that the values in the table satisfy a cubic equation $f(x) = ax^3 + bx^2 + cx + d$ and find a, b, c, and d. Locate the real root of $f(x) = 0$ and find it correct to six decimal places.

x	1	2	3	4	5	6
$f(x)$	5	2	11	38	89	170

(A.E.B.)

8. Show, by means of a sketch, that $6 \sin x = x^3$ has one positive root only and evaluate it correct to four significant figures.

(A.E.B.)

9. Tabulate $f(x) \equiv 3 \cdot 280x^3 + 11 \cdot 827x^2 + 58 \cdot 712\,x + 65 \cdot 125$ for $x = -2 \cdot 0\,(0 \cdot 1) - 1 \cdot 0$ to 5D.

By inverse interpolation, or otherwise, evaluate to four decimal places the root of $f(x) = 0$ in this range.

10. Find the positive root of the equation $\log_e (1 + x) = 2x(x - 1)$ correct to 6D.

11. Show graphically that the equation $e^{2x^2} = 16/x$ has a positive root in the range $0 < x < 2$.

By the method of inverse interpolation, or otherwise, evaluate this root correct to 7S.

(A.E.B.)

12. The differential coefficients of a polynomial are tabulated below. Find a value of x, in the given range, for which the polynomial has a turning point. Express your answer to 6D.

x	0·0	0·1	0·2	0·3	0·4	0·5
$\dfrac{dy}{dx}$	5·000	4·408	2·728	0·248	−2·552	−5·000

x	0·6	0·7	0·8	0·9	1·0
$\dfrac{dy}{dx}$	−6·232	−5·192	−0·632	8·888	25·000

(A.E.B.)

13. The following table of exact values of a quartic in x contains a mistake. Find the mistake and correct it.

Using inverse interpolation, find, correct to three decimal places, the root of the equation $f(x) = 10$ which lies between 0·7 and 0·8.

x	0·0	0·1	0·2	0·3	0·4
$f(x)$	1·0000	1·4641	2·0736	2·8561	3·8461
x	0·5	0·6	0·7		
$f(x)$	5·0625	6·5536	8·3521		(A.E.B.)

14. Find the smallest positive root of the equation

$$x^x + 2x - 6 = 0$$

correct to five significant figures.   (A.E.B.)

15. Evaluate the root of $e^{-3x} \sin 2(x + 1) = 4$ in the neighbourhood of $x = -0.5$ and express your answer to five decimal places.
                    (A.E.B.)

16. If $\rho \log_e x + (1 + x) = 0$, complete the following table:

ρ	0·6	0·7	0·8	0·9	1·0
x	0·1477		0·2181		0·2785

                    (A.E.B.)

17. Show, graphically, that the numerically smallest root of equation

$$x^3 + 4x^2 - 32x = 8 \cosh \frac{x}{2}$$

is in the neighbourhood of -0.2. Calculate this root correct to five decimal places.
                    (A.E.B.)

3
Lagrange Interpolation

3.1 INTRODUCTION

In the previous two chapters we have been concerned with interpolation formulae which have made use of difference tables and various orders of differences. In this chapter we consider interpolation formulae which make use of function values only, together with their corresponding values of the independent variable. Such formulae which may be used with either equal or unequal intervals of the independent variable are known as Lagrange interpolation formulae. These formulae, with the appropriate forms of the coefficients, were first published by Lagrange in 1796, in his *Leçons elementaires sur les mathematiques.*

If the tabulation is at equal intervals, in most cases a central difference formula is to be preferred. In any case, it can be shown, that, at equal intervals, the Lagrange formulae are in fact algebraic rearrangements of the finite difference formulae.

We do not derive the Lagrange formula here but state it and justify its use and accuracy through experience and from the results obtained from its practical application.

The Lagrange formula will be of the form

$$f(x) = \sum_{i=0}^{n} L_i(x) . f_i \qquad (1)$$

where the summation is carried out over the $(n + 1)$ points $x_0, x_1, x_2, x_3, \ldots, x_n$ and $L_i(x)$ are the Lagrangian interpolation coefficients, evaluated at the appropriate point x.

As will be seen an $(n + 1)$ point formula, will in general lead to a polynomial of degree 'n', which is satisfied by the $(n + 1)$ successive values of the function. The required interpolate is then obtained by evaluating the polynomial for the required value of x.

3.2 LAGRANGE COEFFICIENTS

We require a polynomial which is satisfied by the $(n + 1)$ function values at corresponding values of the independent variable. Also we expect this polynomial to be of degree n.

We define the Lagrange interpolation coefficients

$L_i(x)$ for $i = 0, 1, 2, \ldots, n$;

$$L_i(x) \equiv \frac{(x - x_0)(x - x_1) \ldots (x - x_{i-1})(x - x_{i+1})(x - x_{i+2}) \ldots (x - x_n)}{(x_i - x_0)(x_i - x_1) \ldots (x_i - x_{i-1})(x_i - x_{i+1})(x_i - x_{i+2}) \ldots (x_i - x_n)} \quad (2)$$

where the numerator is the product of n factors of the form $(x - x_r)$ in which $(x - x_i)$ has been omitted, and the denominator is the value of this product at $x = x_i$.

Thus there will be $(n + 1)$ such coefficients from the $(n + 1)$ points each giving a polynomial in x of degree n, so that their sum will give a polynomial of degree less than or equal to n.

For on expanding (1) and expressing it in a convenient notation, we have the Lagrange polynomial in the form

$$L(x) \equiv L_0(x).f_0 + L_1(x).f_1 + L_2(x).f_2 + \ldots + L_n(x).f_n \quad (3)$$

where the right-hand side combines $(n + 1)$ polynomials in x, each of degree n. In fact, the Lagrange polynomial for $(n + 1)$ points will be of degree n unless we have the special circumstances existing where the $(n + 1)$ points (x_i, f_i) are exact and satisfy a function of x of degree less than n. This latter case is considered in the next section 3.3.

Further considering (2) it can be easily seen that

$$L_i(x_j) = 0 \qquad \text{if } i \neq j \quad (4)$$

because one term of the numerator will be zero.

Also

$$L_i(x_i) = 1 \quad (5)$$

Using these two results and substituting for the general case $x = x_i$ in (3) we have

$$L(x_i) = 0 .f_0 + 0 .f_1 + 0 .f_2 + \ldots + 1 .f_i + \ldots + 0 .f_n$$

giving

$$L(x_i) = f_i \text{ for all } i = 0, 1, 2, \ldots, n.$$

Thereby proving, that if the Lagrange coefficients are as given by (1) and the Lagrange polynomial is of form (3) this polynomial is satisfied by the $(n + 1)$ points $(x_0, f_0); (x_1, f_1); \ldots (x_n, f_n)$.

3.3 SPECIAL CASE OF $L(x) \equiv f(x)$ FOR $f(x)$ OF DEGREE $m < n$

We now consider the special case mentioned in the previous section, when the given $(n + 1)$ function values are exact and obtained from a polynomial of degree 'm' less than 'n'. Under these conditions we show that the Lagrangian interpolation polynomial $L(x)$ is identical with $f(x)$, the function from which the tabular values are calculated.

Consider the function

$$\frac{f(x)}{(x - x_0)(x - x_1) \ldots (x - x_n)}$$

which we will expand by partial fractions, where $x_0, x_1, \ldots, x_n$, are the $(n + 1)$ values of the independent variable given.

i.e.

$$\frac{f(x)}{(x - x_0)(x - x_1) \ldots (x - x_n)} \equiv \frac{A_0}{(x - x_0)} + \frac{A_1}{x - x_1} + \frac{A_2}{x - x_2} + \ldots$$

$$+ \frac{A_n}{x - x_n}$$

where $A_0, \ldots A_n$, are constants.

$$f(x) \equiv A_0(x - x_1)(x - x_2) \ldots (x - x_n) + A_1(x - x_0)(x - x_2) \ldots (x - x_n) +$$

$$\ldots + A_n(x - x_0)(x - x_1) \ldots (x - x_{n-1})$$

substituting $x = x_0$ we have

$$f(x_0) = A_0(x_0 - x_1)(x_0 - x_2) \ldots (x_0 - x_n)$$

similarly $f(x_1) = A_1(x_1 - x_0)(x_1 - x_2) \ldots (x_1 - x_n)$

$$f(x_2) = A_2(x_2 - x_0)(x_2 - x_1)(x_2 - x_3) \ldots (x_2 - x_n)$$

$$\vdots = \vdots$$

$$f(x_n) = A_n(x_n - x_0)(x_n - x_1)(x_n - x_2) \ldots (x_n - x_{n-1})$$

from which we may obtain expressions for $A_0, A_1, \ldots, A_n$ and substituting for these gives

$$\frac{f(x)}{(x - x_0)(x - x_1) \ldots (x - x_n)} \equiv \frac{f_0}{(x_0 - x_1)(x_0 - x_2) \ldots (x_0 - x_n)} \cdot \frac{1}{x - x_0}$$

$$+ \frac{f_1}{(x_1 - x_0)(x_1 - x_2) \ldots (x_1 - x_n)} \cdot \frac{1}{x - x_1} + \ldots$$

$$+ \frac{f_n}{(x_n - x_0)(x_n - x_1) \ldots (x_n - x_{n-1})} \cdot \frac{1}{x - x_n}$$

which gives

$$f(x) \equiv \frac{(x - x_1)(x - x_2) \ldots (x - x_n)}{(x_0 - x_1)(x_0 - x_2) \ldots (x_0 - x_n)} \cdot f_0$$

$$+ \frac{(x - x_0)(x - x_2) \ldots (x - x_n)}{(x_1 - x_0)(x_1 - x_2) \ldots (x_1 - x_n)} \cdot f_1$$

$$+ \ldots$$

$$+ \frac{(x - x_0)(x - x_1) \ldots (x - x_{n-1})}{(x_n - x_0)(x_n - x_1) \ldots (x_n - x_{n-1})} \cdot f_n$$

$$\equiv L(x) \text{ as given in (3).}$$

Hence if $f(x) \equiv L(x)$, in this special case, where $f(x)$ is of degree m (less than n), then in the Lagrange expansion of the right-hand side the terms in x^{m+1}, $x^{m+2}, \ldots, x^n$ will have zero coefficients. This means that the various terms in $x^{m+1}, x^{m+2}, \ldots, x^n$, which occur in the separate terms of the right-hand side all cancel when they are combined together.

3.3.1 Condition connecting Lagrange coefficients

We consider the special case of § 3.3 where the degree of the Lagrange polynomial exceeds the degree of the function from which the function values are evaluated. We use the above result to deduce a particular relationship connecting the Lagrange coefficients $L_i(x)$.

In particular let $f(x) \equiv c$ (a constant) then the given $(n + 1)$ points with co-ordinates $\{x_n, f(x_n)\}$; $n = 0, 1, 2, \ldots, n$ take the form

$$(x_0, c); (x_1, c); (x_2, c); (x_3, c); \ldots (x_n, c)$$

which on substituting in (3) gives

$$L(x) = L_0(x) \cdot c + L_1(x) \cdot c + L_2(x) \cdot c + \ldots + L_n(x) \cdot c$$

but the given function $f(x)$ has degree less than n and using the conclusion of the previous section

$$L(x) \equiv f(x) = c$$

∴ $$c = c\{L_0(x) + L_1(x) + L_2(x) + \ldots + L_n(x)\}$$

giving $$1 = L_0(x) + L_1(x) + L_2(x) + \ldots + L_n(x) \qquad (6)$$

An important result showing that the *sum of the Lagrange coefficients is one*.

We will frequently make use of this result to check the accuracy of future calculations dealing with Lagrange interpolation. This result is true for any function values and any values of the independent variable x, i.e. $x_0, x_1, \ldots, x_n$ because the coefficients as given in (2) depend only on the values $x_0, x_1, \ldots x_n$. The result (6) means that when the Lagrange coefficients are added together the terms in x^r, for all $r > 0$, will cancel out. We demonstrate this in the next example.

3.3.2 Worked example to show that the sum of the Lagrangian interpolation coefficients add up to one.

Choosing three arbitrary values of x_n as

$$x_0 = -2, x_1 = 2, x_2 = 6.$$

We have using (2)

$$L_0(x) = \frac{(x - 2)(x - 6)}{(-2 - 2)(-2 - 6)} \Rightarrow L_0(x) = \frac{x^2 - 8x + 12}{32}$$

$$L_1(x) = \frac{(x + 2)(x - 6)}{(2 + 2)(2 - 6)} \Rightarrow L_1(x) = \frac{x^2 - 4x - 12}{-16}$$

$$L_2(x) = \frac{(x+2)(x-2)}{(6+2)(6-2)} \quad \Rightarrow L_2(x) = \frac{x^2-4}{32}$$

$$\therefore \sum_{i=0}^{2} L_i(x) = \frac{x^2-8x+12}{32} + \frac{(-)(x^2-4x-12)}{16} + \frac{x^2-4}{32}$$

$$= \frac{x^2-8x+12-2x^2+8x+24+x^2-4}{32}$$

$$= 1.$$

i.e. Sum of the Lagrange coefficients is 1.

3.4   SECOND DEGREE LAGRANGE INTERPOLATION

Suppose now that we are given corresponding function values for values of the independent variable.

e.g. $\{x_0 = -2, \quad f_0 = -8{\cdot}52\}$; $\{x_1 = 1, \quad f_1 = 3{\cdot}00\}$;

$$\{x_2 = 2, \quad f_2 = 7{\cdot}48\}.$$

We may proceed as above to obtain the Lagrangian interpolation coefficients as functions of x.

Using (2) we have

$$L_0(x) \equiv \frac{(x-1)(x-2)}{(-2-1)(-2-2)} \qquad \text{gives } L_0(x) \equiv \frac{x^2-3x+2}{12}$$

$$L_1(x) \equiv \frac{(x+2)(x-2)}{(1+2)(1-2)} \qquad \text{gives } L_1(x) \equiv \frac{x^2-4}{-3}$$

$$L_2(x) \equiv \frac{(x+2)(x-1)}{(2+2)(2-1)} \qquad \text{gives } L_2(x) \equiv \frac{x^2+x-2}{4}$$

and adding these Lagrange coefficients

$$\sum_{0}^{2} L_i(x) = \frac{x^2-3x+2}{12} + \frac{x^2-4}{-3} + \frac{x^2+x-2}{4}$$

$$= \frac{x^2-3x+2+(-4)(x^2-4)+3(x^2+x-2)}{12}$$

$$= 1$$

again satisfying the condition for the coefficients and suggesting that the working is accurate.

Now using the Lagrangian Interpolation Polynomial as given by

$$L(x) \equiv \sum L_i(x) . f_i$$

$$L(x) \equiv \frac{-8 \cdot 52(x^2 - 3x + 2)}{12} + \frac{3 \cdot 00(x^2 - 4)}{-3} + \frac{7 \cdot 48(x^2 + x - 2)}{4}$$

$$\equiv -0 \cdot 71(x^2 - 3x + 2) - x^2 + 4 + 1 \cdot 87(x^2 + x - 2)$$

$$\therefore \quad L(x) \equiv 0 \cdot 16x^2 + 4x - 1 \cdot 16$$

Whence using a machine and the method of 'nested multiplication' we have

$$L(1 \cdot 5) = 5 \cdot 20.$$

This is an example of a *Second Degree Lagrange Interpolation*.

3.5 EXAMPLES

1. Given the following values of $x = 1$, 3 and 6 find the Lagrange interpolation coefficients and verify that the sum of the coefficients is unity.

2. Evaluate the Lagrange interpolation coefficients corresponding to $x_0 = -1$, $x_1 = 1$, and $x_2 = 2$. Verify that the sum of the coefficients is unity.

3. Evaluate the Lagrange interpolation coefficients given

$x_0 = -3, x_1 = 0, x_2 = 2, x_3 = 5$. Verify that $\Sigma L_i(x) = 1$.

4. Given the data

x	0	1	3
$f(x)$	-3	-5	18

use the method of section 3.4 to form the second-degree Lagrange interpolation polynomial. Hence find the value of $f(x)$ when $x = 2$.

5. Obtain a second-degree Lagrange interpolation formula from the following information:

x	-2	0	1
$f(x)$	$10 \cdot 75$	$-1 \cdot 65$	$1 \cdot 45$

and hence find $f(0 \cdot 5)$ assuming the function values are exact values of a second degree polynomial.

6. The following four function values and corresponding values of the independent variable are exact values of the function:

x	-0.5	1.5	2	3
$f(x)$	-1.0	1.0	2.75	7.75

Given that they are obtained from a quadratic function use three of these
values to find the quadratic and use the fourth as a check.

3.6   ERRORS IN LAGRANGE INTERPOLATION

Before proceeding to higher order Lagrange interpolation we consider the
problem of errors and how they develop in the method of interpolation. We
have already mentioned that if the given tabular values are at equal intervals
then central difference formulae are to be preferred, as in Chapter 1.

However for tabular values at *unequal intervals, only Lagrange interpolation
can be used* and consequently no table of differences is required. This is an
advantage but Lagrange interpolation does have disadvantages.

3.6.1   Disadvantages of Lagrange interpolation

(i) When using the function values to obtain the interpolation polynomial
there is no check on their accuracy as is automatically supplied when a 'Table
of differences' is formed for use with central difference formulae.

(ii) When a Lagrange polynomial is used to obtain an interpolate it is impossible to determine whether the degree of the polynomial used is correct.
Only after several values of the interpolate have been obtained using Lagrange
formulae of different degrees, the answers compared and the number of common figures checked, can we be sure that a formula of sufficiently high order has
been used. If extra function values are available this affords a convenient method
of checking accuracy. We demonstrate this in § 3.7.3 and §3.7.4.

(iii) Lagrange formulae do not lend themselves to easy arrangement for inverse interpolation.

3.6.2.   Sources of errors

(i) As with finite difference formulae we will have rounding-off errors in the
given function values. Then, because each function value may contain a rounding off error, when using the formula

$$L(x) = \sum_{0}^{n} L_r(x) \cdot f_r$$

these errors may accumulate. Thus it is advisable to carry at least one guarding figure when evaluating the Lagrange coefficients.

(ii) Truncation errors occur through the truncation of the polynomial, which in the case of Lagrange interpolation happens when fewer points are used, so giving different degree polynomials. As mentioned in § 3.6.1 only after several values of the interpolate are calculated can a decision be made about its accuracy. In practice, if an $(n + 1)$ point formula is used first and no extra function values are available the interpolation should be repeated with n function values and the figures common in the two interpolates give a measure of the accuracy possible from the given data. This is demonstrated in § 3.7.5.

3.7 HIGHER-ORDER LAGRANGIAN INTERPOLATION
THIRD-ORDER INTERPOLATION

It is an important consideration at this stage to reduce the numerical work to a minimum. In this section we consider two methods of solving the given problem.

Consider the following given function values and their corresponding value of x.

	x_0	x_1	x_2	x_3
x	-4	-1	0	2
$f(x)$	-26	4	2	16

Evaluate $f(1 \cdot 2)$ correct to 2D.

3.7.1 Method A

In this method we incorporate one positive check. Other checks occur incidentally because of the symmetry of the scheme.

Further we consider now the possibility of errors. Since $\Sigma |f(x)| = 48$ when we calculate $\Sigma L_i(x) \cdot f_i(x)$ there may be an accumulated error, which will affect the accuracy of the interpolate. For if the $L_i(x)$ are calculated to 4D there is a possible error of $\pm \frac{1}{2} \times 10^{-4}$ in each and if this is multiplied by 48 gives $\pm 24 \times 10^{-4}$ or $\pm 0 \cdot 0024$ as maximum error. In practice this is unlikely to be the case and we can confidently expect two decimal place accuracy, working with $L_r(x)$ to 4D.

Here is a suggested method of lay-out for this type of example.

The first three columns should be completed first, before the column $L_r(x_p)$ is attempted. Then using (2) of § 3.2 one should note that the numerator of each $L_r(x_p)$ uses the quantities in the $(x_p - x_r)$ column excluding the one in the same row. We show this in the case of $L_0(x_p)$ with the arrows showing the quantities in the numerator.

r	x_r	$x_p - x_r$	$L_r(x_p)$	L_r	f_r	$L_r \cdot f_r$
0	-4	$5\cdot2$	$\dfrac{2\cdot2 \times 1\cdot2 \times -0\cdot8}{-3 \times -4 \times -6}$ $= 0\cdot0293$		-26	$-0\cdot7618$
1	-1	$2\cdot2$	$\dfrac{5\cdot2 \times 1\cdot2 \times -0\cdot8}{3 \times -1 \times -3}$ $= -0\cdot5547$		4	$-2\cdot2188$
2	0	$1\cdot2$	$\dfrac{5\cdot2 \times 2\cdot2 \times -0\cdot8}{4 \times 1 \times -2}$ $= 1\cdot144$		2	$2\cdot288$
3	2	$-0\cdot8$	$\dfrac{5\cdot2 \times 2\cdot2 \times 1\cdot2}{6 \times 3 \times 2}$ $= 0\cdot3813$		16	$6\cdot1008$

Check $\Sigma\, L_r = 0\cdot9999$ $L(x) = 5\cdot4082$

The quantities $L_r(x_p)$ are worked out on the machine and then $\Sigma L_r(x_p)$ is found and checked that it gives one. In this particular example the slight error occurs because of the rounding-off of the individual $L_r(x_p)$.

Adding the last column gives

$$f(1\cdot2) = 5\cdot41 \text{ (2D)}$$

3.7.2   Method B

In this method we first establish the third-order Lagrange interpolation polynomial and then find $f(1\cdot2)$ by direct substitution. Following the method of § 3.3.2

$$L_0(x) = \frac{(x + 1)(x)(x - 2)}{-3 \times -4 \times 6} \qquad \text{giving } L_0 = \frac{x^3 - x^2 - 2x}{-72}$$

$$L_1(x) = \frac{(x + 4)(x)(x - 2)}{3 \times -1 \times -3} \qquad \text{giving } L_1 = \frac{x^3 + 2x^2 - 8x}{9}$$

$$L_2(x) = \frac{(x + 4)(x + 1)(x - 2)}{4 \times 1 \times -2} \qquad \text{giving } L_2 = \frac{x^3 + 3x^2 - 6x - 8}{-8}$$

$$L_3(x) = \frac{(x + 4)(x + 1)x}{6 \times 3 \times 2} \qquad \text{giving } L_3 = \frac{x^3 + 5x^2 + 4x}{36}$$

again we show that $\Sigma L_r(x) = 1$ as a check.

for $\Sigma L_r(x) = \dfrac{-(x^3 - x^2 - 2x) + 8(x^3 + 2x^2 - 8x)}{72}$

$\qquad - \dfrac{9(x^3 + 3x^2 - 6x - 8) + 2(x^3 + 5x^2 + 4x)}{72} = 1$ by inspection

Hence $L(x) = \Sigma L_r(x) \cdot f_r$ gives

$L(x) \equiv \dfrac{-26(x^3 - x^2 - 2x)}{-72} + \dfrac{4(x^3 + 2x^2 - 8x)}{9} + \dfrac{2(x^3 + 3x^2 - 6x - 8)}{-8}$

$\qquad\qquad\qquad\qquad\qquad + \dfrac{16(x^3 + 5x^2 + 4x)}{36}$

$L(x) \equiv x^3 \left(\dfrac{13}{36} + \dfrac{4}{9} - \dfrac{1}{4} + \dfrac{16}{36}\right) + x^2 \left(-\dfrac{13}{36} + \dfrac{8}{9} - \dfrac{3}{4} + \dfrac{80}{36}\right) +$

$\qquad\qquad\qquad\qquad x \left(-\dfrac{26}{36} - \dfrac{32}{9} + \dfrac{3}{2} + \dfrac{64}{36}\right) + 2$

$\therefore L(x) \equiv x^3 + 2x^2 - x + 2$

This is the third-order Lagrange interpolation polynomial obtained from the given data.

Then, by machine we obtain $L(1\cdot2) = 5\cdot408$

$\qquad\qquad$ giving $f(1\cdot2) = 5\cdot41$ (2D) as before

In practice it is found that this second method B will take longer than the first method A, particularly when the values of the independent variable and the given function values involve decimal quantities. However, method B might well be preferable in a case, where from the given data a number of interpolates were to be obtained.

The two methods illustrate the two possible uses of Lagrangian interpolation, namely

(i) finding an interpolate directly (method A),

(ii) finding the interpolation formula first and then using it to find the inter-polates (method B).

3.7.3 A worked example, in which higher-order interpolation is used, is considered next.

$x_0 = 2\cdot3$ $\qquad$ $x_1 = 2\cdot4$ $\qquad$ $x_2 = 2\cdot6$ $\qquad$ $x_3 = 2\cdot7$

$f_0 = 0\cdot149\ 97$ $\quad$ $f_1 = 1\cdot382\ 84$ $\quad$ $f_2 = 4\cdot166\ 76$ $\quad$ $f_3 = 5\cdot726\ 33$

$x_4 = 2\cdot8$ $\qquad$ $x_5 = 3\cdot0$

$f_4 = 7\cdot403\ 32$ $\quad$ $f_5 = 11\cdot126\ 60$

First we use a five point Lagrange interpolation to obtain $f(2\cdot423)$.

In this case, as we have been given 6 points, we choose the five points to use so that $x_p = 2\cdot423$ is centrally placed. That is, in this particular example we use the first five given.

As the function values are given to 5D and $\Sigma |f(x)| \simeq 30$, for the same reasons as in §3.7.1 we evaluate the Lagrange coefficients to two extra decimal places, that is to 7D.

Using method A of previous example:

r	x_r	$x_p - x_r$	$L_r(x_p)$	L_r	f_r	$L_r \cdot f_r$
0	2·3	0·123	$\dfrac{0\cdot023 \times -0\cdot177 \times -0\cdot277 \times -0\cdot377}{0\cdot1 \times -0\cdot3 \times -0\cdot4 \times -0\cdot5}$	$= -0\cdot070\ 855\ 1$	0·149 97	$-0\cdot010\ 626\ 1$
1	2·4	0·023	$\dfrac{0\cdot123 \times -0\cdot177 \times -0\cdot277 \times -0\cdot377}{0\cdot1 \times -0\cdot2 \times -0\cdot3 \times -0\cdot4}$	$= 0\cdot947\ 301\ 7$	1·382 84	1·309 966 7
2	2·6	−0·177	$\dfrac{0\cdot123 \times 0\cdot023 \times -0\cdot277 \times -0\cdot377}{0\cdot3 \times 0\cdot2 \times -0\cdot1 \times -0\cdot2}$	$= 0\cdot246\ 191\ 4$	4·166 76	1·025 820 5
3	2·7	−0·277	$\dfrac{0\cdot123 \times 0\cdot023 \times -0\cdot177 \times -0\cdot377}{0\cdot4 \times 0\cdot3 \times 0\cdot1 \times -0\cdot1}$	$= -0\cdot157\ 313\ 6$	5·726 33	$-0\cdot900\ 829\ 6$
4	2·8	−0·377	$\dfrac{0\cdot123 \times 0\cdot023 \times -0\cdot177 \times -0\cdot277}{0\cdot5 \times 0\cdot4 \times 0\cdot2 \times 0\cdot1}$	$= 0\cdot034\ 675\ 8$	7·403 32	0·256 716 0

Check $\Sigma L_r = 1\cdot000\ 000\ 2$ $L(x) = 1\cdot681\ 047\ 5$

All the calculations performed by machine, the last column

$$L(x) = \sum_{0}^{4} L_r(x_p) \cdot f_r$$

gives $f(2{\cdot}423) = 1{\cdot}681\ 047\ 5$

$\qquad\qquad = 1{\cdot}681\ 05\ (5D)$

3.7.4 As a check, bearing in mind that we have an extra point available, we will repeat the evaluation of the interpolate using a six-point formula. Evaluating the coefficients as before, to seven decimal places, give

$L_0(x_p) = -\ 0{\cdot}058\ 404\ 8 \qquad L_1(x_p) = \quad 0{\cdot}910\ 988\ 5$

$L_2(x_p) = \quad 0{\cdot}355\ 131\ 1 \qquad L_3(x_p) = -\ 0{\cdot}302\ 566\ 5$

$L_4(x_p) = \quad 0{\cdot}100\ 039\ 7 \qquad L_5(x_p) = -\ 0{\cdot}005\ 187\ 6$

giving check $\Sigma L_r(x_p) = 1{\cdot}000\ 000\ 4$ and so assuming that these coefficients are correct, using the six-point summation $L(x) = \sum_{0}^{5} L_r(x) \cdot f_r$ we have

$L(2{\cdot}423) = 1{\cdot}681\ 048\ 4$ or once again

$f(2{\cdot}423) = 1{\cdot}681\ 05\ (5D)$

3.7.5 The only alternative method of performing a check, had the sixth point not been available, would have been to omit one of the original five points used and perform a 4-point Lagrange interpolation. Such considerations help to show, how in general, there is uncertainty in the degree of interpolation to be used in Lagrange methods.

Using the first four points, because they are spaced, two on either side of the $x_p = 2{\cdot}423$, the Lagrange coefficients for the four-point interpolation are

$L_0(x_p) = -\ 0{\cdot}093\ 972\ 3 \qquad L_1(x_p) = \quad 1{\cdot}005\ 094\ 5$

$L_2(x_p) = +\ 0{\cdot}130\ 605\ 5 \qquad L_3(x_p) = -\ 0{\cdot}041\ 727\ 8$

where the check of $\Sigma L_r(x_p) = 0{\cdot}999\ 999\ 9$. Giving $L(2{\cdot}423) = \sum_{0}^{3} L_t(x_p) \cdot f_t$

$\qquad\qquad = 1{\cdot}681\ 046\ 5$

$\qquad\qquad = 1{\cdot}681\ 05\ (5D)$

3.7.6 As a further check, still using the initial data as given at the start of § 3.7.3 we conclude by finding the Langrange polynomial of degree 3; we use the first four points as follows

$x_0 \quad 2{\cdot}3 \quad \dfrac{(x - 2{\cdot}4)\ (x - 2{\cdot}6)\ (x - 2{\cdot}7)}{(-\ 0{\cdot}1) \times (-\ 0{\cdot}3) \times (-\ 0{\cdot}4)} \Rightarrow \dfrac{x^3 - 7{\cdot}7x^2 + 19{\cdot}74x - 16{\cdot}848}{-\ 0{\cdot}012}$

$x_1 \quad 2{\cdot}4 \quad \dfrac{(x - 2{\cdot}3)\ (x - 2{\cdot}6)\ (x - 2{\cdot}7)}{(0{\cdot}1)\ (-\ 0{\cdot}2)\ (-\ 0{\cdot}3)} \Rightarrow \dfrac{x^3 - 7{\cdot}6x^2 - 19{\cdot}21x - 16{\cdot}146}{0{\cdot}006}$

$$x_2 \quad 2{\cdot}6 \quad \frac{(x - 2{\cdot}3)(x - 2{\cdot}4)(x - 2{\cdot}7)}{(0{\cdot}3)(0{\cdot}2)(-0{\cdot}1)} \Rightarrow \frac{x^3 - 7{\cdot}4x^2 + 18{\cdot}21x - 14{\cdot}904}{-0{\cdot}006}$$

$$x_3 \quad 2{\cdot}7 \quad \frac{(x - 2{\cdot}3)(x - 2{\cdot}4)(x - 2{\cdot}6)}{(0{\cdot}4)(0{\cdot}3)(0{\cdot}1)} \Rightarrow \frac{x^3 - 7{\cdot}3x^2 + 17{\cdot}74x - 14{\cdot}352}{0{\cdot}012}$$

Then using $L(x) \equiv \overset{3}{\underset{0}{\Sigma}} L_r(x) \cdot f_r$ gives

$$L(x) \equiv \left(\frac{x^3 - 7{\cdot}7x^2 + 19{\cdot}74x - 16{\cdot}848}{-0{\cdot}012} \right)(0{\cdot}149\ 97)$$

$$+ \left(\frac{x^3 - 7{\cdot}6x^2 + 19{\cdot}21x - 16{\cdot}146}{0{\cdot}006} \right)(1{\cdot}382\ 84)$$

$$+ \left(\frac{x^3 - 7{\cdot}4x^2 + 18{\cdot}21x - 14{\cdot}904}{-0{\cdot}006} \right)(4{\cdot}166\ 76)$$

$$+ \left(\frac{x^3 - 7{\cdot}3x^2 + 17{\cdot}74x - 14{\cdot}352}{0{\cdot}012} \right)(5{\cdot}726\ 33)$$

from which, on rounding the coefficients to 4D we have

$$L(x) \equiv 0{\cdot}71x^3 + 0{\cdot}12x^2 - 9{\cdot}1234$$

Using this Lagrange polynomial to find the interpolate for $x = 2{\cdot}423$ gives $L(2{\cdot}423) = 1{\cdot}681\ 046\ 8$ or as before

$$L(2{\cdot}423) = 1{\cdot}681\ 05 \ (5D).$$

3.7.7 Conclusion. If we compare the results of the four solutions obtained there is obviously close agreement. We conclude that probably $L(2{\cdot}423)$ is equal to $1{\cdot}681\ 05$ (5D). The examples show clearly how difficult it is to assess from the given data how many points to use in the interpolation, a situation which is equally true for data given at equal intervals. Whereas in equal intervals, using central difference formulae the difference table gives a lead as to where the formula should be truncated, this is not possible with Lagrangian interpolation. Here, each change in order means restarting and recalculating the coefficients, as we have done. It would be useful practice to check the calculations in sections § 3.7.4 and § 3.7.5.

3.8 AITKEN'S METHOD FOR LAGRANGE INTERPOLATION AT UNEQUAL INTERVALS

We conclude by introducing a method of interpolation developed by Aitken, which goes a long way in overcoming the disadvantage of not knowing how many points to use for any one particular interpolation. In this method successive interpolates of order 1, 2, 3, . . . are formed systematically, each being obtained by a method basically similar to linear interpolation. For this reason, the method is sometimes referred to as 'interpolation by a sequence of linear cross-means'.

As an introduction to this theory, consider the case where we are given three function values and their associated values of the independent variable. Let these be $A(x_0, f_0)$; $B(x_1, f_1)$; $C(x_2, f_2)$ and x as the value within the range, such that $x_1 < x < x_2$ and at which the interpolate is required. We plot these points on a diagram and a function curve through them.

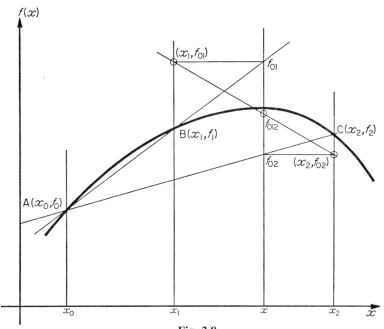

Fig. 3.8

The chord AB is produced to cut the ordinate at x, to give point f_{01} and similarly the chord AC for the point f_{02}.

Theoretically, the equation of the chord AB is

$$y - f_0 = \frac{f_1 - f_0}{x_1 - x_0}(x - x_0) \qquad \text{(ref § 1.1.1)}$$

giving

$$y(x_1 - x_0) - (x_1 - x_0)f_0 = (f_1 - f_0)(x - x_0)$$

$$\therefore y(x_1 - x_0) = (x_1 - x_0 - x + x_0)f_0 + (x - x_0)f_1$$

$$\therefore y = \frac{(x_1 - x)f_0 + (x - x_0)f_1}{x_1 - x_0}$$

or alternatively in a more convenient form as

$$y = \frac{(x - x_0)f_1 - (x - x_1)f_0}{x_1 - x_0} \qquad (7)$$

for those familiar with determinants, (7) can be expressed in form

$$y = \frac{\begin{vmatrix} x - x_0 & f_0 \\ x - x_1 & f_1 \end{vmatrix}}{(x_1 - x_0)} \tag{8}$$

This is the most convenient form. Aitken calls the expression given in (7) the 'linear cross-mean' between f_0 and f_1 and in notation it is written as $f_{01}(x)$.

Thus, by a similar method

$$f_{02}(x) = \frac{(x - x_0)f_2 - (x - x_2)f_0}{(x_2 - x_0)} \quad \text{or} \quad \frac{\begin{vmatrix} x - x_0 & f_0 \\ x - x_2 & f_2 \end{vmatrix}}{(x_2 - x_0)}$$

and more generally

$$f_{0i} = \frac{(x - x_0)f_i - (x - x_i)f_0}{x_i - x_0} \quad \text{or} \quad \frac{\begin{vmatrix} x - x_0 & f_0 \\ x - x_i & f_i \end{vmatrix}}{x_i - x_0} \tag{9}$$

for $i = 1, 2, 3, \ldots, n$.

Note that $f_{0i}(x_0)$ gives on substitution into the determinant form of (9)

$$f_{0i}(x_0) = \frac{\begin{vmatrix} (x_0 - x_0) & f_0 \\ (x_0 - x_i) & f_i \end{vmatrix}}{(x_i - x_0)}$$

$$= \frac{-(x_0 - x_i) \cdot f_0}{(x_i - x_0)}$$

$$= f_0$$

and more generally

$$f_{0i}(x_i) = f_i \tag{10}$$

as is required by any interpolation formula.

We continue the process of forming linear cross-means to obtain the quantity $f_{012}(x)$ by using the same procedure on the quantities $(x_1, f_{01}(x))$; $(x_2, f_{02}(x))$ shown on the diagram fig. 3.8 by the points marked with small circles.

Then using the non-determinant form of (9) we have

$$f_{012}(x) = \frac{(x - x_1)f_{02}(x) - (x - x_2)f_{01}(x)}{x_2 - x_1}$$

or in determinant form as

$$f_{012}(x) = \frac{1}{x_2 - x_1} \begin{vmatrix} x - x_1 & f_{01}(x) \\ x - x_2 & f_{02}(x) \end{vmatrix}$$

generalised as

$$f_{01i}(x) = \frac{(x - x_1)f_{0i}(x) - (x - x_i)f_{01}(x)}{x_i - x_1}$$

or in determinant form

$$f_{01i}(x) = \frac{1}{x_i - x_1} \begin{vmatrix} x - x_1 & f_{01}(x) \\ x - x_i & f_{0i}(x) \end{vmatrix} \tag{11}$$

We now form a table showing the results obtained so far which shows clearly the point of using a determinant notation. Note also in the table, the second column consisting of quantities of the form $x - x_i$, which following Aitken are called 'PARTS'.

PARTS

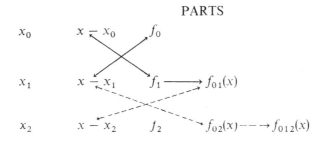

the diagonal lines showing the position of two of the determinant forms involved.

3.8.1 Before considering a numerical example, we show that the quantity $f_{012}(x)$ is in fact, a second order Lagrange interpolation polynomial for the given data. We have

$$f_{012}(x) \equiv \frac{(x - x_1) \cdot f_{02}(x) - (x - x_2) \cdot f_{01}(x)}{(x_2 - x_1)}$$

which, on substituting for $f_{02}(x)$ and $f_{01}(x)$ from (11) gives

$$f_{012}(x) \equiv$$

$$\frac{(x - x_1) \left\{ \dfrac{(x - x_0) f_2 - (x - x_2) f_0}{x_2 - x_0} \right\} - (x - x_2) \left\{ \dfrac{(x - x_0) f_1 - (x - x_1) f_0}{(x_1 - x_0)} \right\}}{(x_2 - x_1)}$$

$$= \frac{(x - x_1)(x_1 - x_0) \{ (x_1 - x_0) \cdot f_2 - (x - x_2) f_0 \}}{(x_1 - x_0)(x_2 - x_0)(x_2 - x_1)}$$

$$\frac{- (x - x_2)(x_2 - x_0)\{ (x - x_0) \cdot f_1 - (x - x_1) f_0 \}}{(x_1 - x_0)(x_2 - x_0)(x_2 - x_1)}$$

Considering only the term in f_0 we have

$$f_0[(x - x_2)(x_2 - x_0)(x - x_1) - (x_1 - x_0)(x - x_1)(x - x_2)]$$

$$= (x - x_1)(x - x_2) \cdot f_0 \{ x_2 - x_0 - x_1 + x_0 \}$$

$$= (x - x_1)(x - x_2)(x_2 - x_1) \cdot f_0 \text{ and on substituting we have}$$

$$f_{012}(x) \equiv \frac{(x - x_1)(x - x_2)(x_2 - x_1) \cdot f_0 - (x - x_0)(x - x_2)(x_2 - x_0) \cdot f_1}{(x_1 - x_0)(x_2 - x_0)(x_2 - x_1)} +$$

$$\frac{(x - x_0)(x - x_1)(x_1 - x_0) \cdot f_2}{(x_1 - x_0)(x_2 - x_0)(x_2 - x_1)}$$

on dividing out

$$f_{012}(x) \equiv \frac{(x - x_1)(x - x_2)}{(x_0 - x_1)(x_0 - x_2)} \cdot f_0 + \frac{(x - x_0)(x - x_2)}{(x_1 - x_0)(x_1 - x_2)} \cdot f_1$$

$$+ \frac{(x - x_0)(x - x_1)}{(x_2 - x_0)(x_2 - x_1)} \cdot f_2$$

which is the Lagrange polynomial of degree 2 as given by (3) in § 3.2. Similarly it could be shown that $f_{013}(x)$ is the Lagrange polynomial of degree 2 obtained from the points (x_0, f_0); (x_1, f_1) and (x_3, f_3).

Thus every quantity of the form $f_{01i}(x)$ is a Lagrange polynomial of degree 2 and in numerical work when these are evaluated, it is possible, by comparison of the number of common significant figures, to assess the accuracy at each stage. Thus, using Aitken's method the table is developed only as far as is required, depending on the accuracy needed in the final interpolate.

3.8.2 As a numerical example consider the one in Section 3.4 where we were given the following data:

$$\{ x_0 = -2, \quad f_0 = -8.52 \}; \{ x_1 = 1, \quad f_1 = 3.00 \} \text{ and } \{ x_2 = 2, \quad f_2 = 7.48 \}$$

we require $f(1.5)$, i.e. $x = 1.5$.

Setting out the table and using a machine

x_r	'Parts' $(x - x_r)$	f_r	f_{0i}	f_{01t}
-2	3.5	-8.52		2nd order
1	0.5	3.00	$f_{01} = 4.92$	Lagrange Interpolate
2	-0.5	7.48	$f_{02} = 5.48$	$f_{012} \overset{\downarrow}{=} 5.20$

The $f_{01}(1.5)$ is obtained, using

$$f_{01}(x) = \frac{(x - x_0)f_1 - (x - x_1)f_0}{(x_1 - x_0)}$$

and evaluating, on substitution

$$\frac{(3.5)(3.00) - (0.5)(-8.52)}{3}$$

$$= 4.92$$

Similarly for the other quantities

Thus, by Aitken's method, $f(1\cdot5) = 5\cdot20$ agreeing exactly with the previous work.

3.8.4 To continue this method further and include higher order Lagrange interpolates we need to consider such quantities as f_{012i}, f_{0123i}, and so on. To do this we develop the table of quantities further to see how such terms are obtained.

x_r	'Parts' $(x - x_r)$	f_r	f_{0i}	f_{01i}	f_{012i}	f_{0123i}	f_{01234i}
x_0	$x - x_0$	f_0					
x_1	$x - x_1$	f_1	$f_{01}(x)$				
x_2	$x - x_2$	f_2	$f_{02}(x)$	$f_{012}(x)$			5th order Lagrange interpolate
x_3	$x - x_3$	f_3	$f_{03}(x)$	$f_{013}(x)$	$f_{0123}(x)$		
x_4	$x - x_4$	f_4	$f_{04}(x)$	$f_{014}(x)$	$f_{0124}(x)$	$f_{01234}(x)$	
x_5	$x - x_5$	f_5	$f_{05}(x)$	$f_{015}(x)$	$f_{0125}(x)$	$f_{01235}(x)$	$f_{012345}(x)$

In particular the quantity $f_{0124}(x)$ is obtained using the points $(x_2, f_{012}(x))$ and $(x_4, f_{014}(x))$ shown in the diagram, to give

$$f_{0124}(x) = \frac{(x - x_2) \cdot f_{014}(x) - (x - x_4) \cdot f_{012}(x)}{x_4 - x_2}$$

and similarly

$$f_{01235}(x) = \frac{(x - x_3) \cdot f_{0125}(x) - (x - x_5) \cdot f_{0123}(x)}{x_5 - x_3}$$

and in the general, for an mth order interpolate

$$f_{0123\,\ldots\,lmn}(x) = \frac{(x - x_m) \cdot f_{0123\,\ldots\,ln}(x) - (x - x_n) \cdot f_{0123\,\ldots\,lm}(x)}{(x_n - x_m)} \tag{12}$$

3.8.5 Further numerical examples using Aitken's method are now considered. First, the example solved in § 3.7.1 and § 3.7.2 using two alternative methods

x	-4	-1	0	2	
$f(x)$	-26	4	2	16	to find $f(1\cdot2)$.

Proceeding as before, entering the first three columns first, then performing the calculations by (12) gives

x_r	'Parts' $(x - x_r)$	$f_r(x)$	f_{0i}	f_{01i}	f_{012i}
-4	5·2	-26			

-1 2·2   4   26   4 point Lagrange interpolation

0   1·2   2   10·4 $\longrightarrow -8{\cdot}32$

| 2 | $-0{\cdot}8$ | 16 | 10·4 | 14·56 | 5·408 |

giving, as before $f(1{\cdot}2) = 5{\cdot}41$ (2D).

With practice, each calculation can be performed as a continuous process on the machine, up to division stage, when one must divide by $(x_n - x_m)$. Complete the entries in the first three columns so that the quantities for the first set of determinants are suitably positioned for the calculations. Generally when we arrive at the stage for division, the quantity $x_n - x_m$ will be obtained mentally, because the intervals will be numerically simple.

3.8.6 As a final example of Aitken's method, we solve the problem given in § 3.7.3 where the given data was

$x_0 = 2{\cdot}3$ $x_1 = 2{\cdot}4$ $x_2 = 2{\cdot}6$ $x_3 = 2{\cdot}7$
$f_0 = 0{\cdot}149\,97$ $f_1 = 1{\cdot}382\,84$ $f_2 = 4{\cdot}166\,76$ $f_3 = 5{\cdot}726\,33$
$x_4 = 2{\cdot}8$ $x_5 = 3{\cdot}0$
$f_4 = 7{\cdot}403\,32$ $f_5 = 11{\cdot}126\,60$ find $f(x)$ when $x = 2{\cdot}423$.

x_r	'Parts' $x - x_r$	f_r	f_{0i}	f_{01i}	f_{012i}
2·3	0·123	0·149 97			
2·4	0·023	1·382 84	1·666 400 1		
2·6	$-0{\cdot}177$	4·166 76	1·796 853 9	1·68|1 402 3	
2·7	$-0{\cdot}277$	5·726 33	1·864 700 7	1·68|1 603 1	104 6|9
2·8	$-0{\cdot}377$	7·403 32	1·934 294 1	1·68|1 804 0	104 6|8
3·0	$-0{\cdot}577$	11·126 60	2·078 720 7	1·68|2 205 7	104 6|8

Having completed the table according to Aitken's method, retaining two extra decimals to ensure minimum rounding-off errors, the best value of $f(2{\cdot}423)$ is 1·681 047 or 1·681 05 (5D) as before.

In particular, notice how the working may be reduced at suitable stages. In the above table all the entries in the f_{01i} column have the same three leading digits, 1·68. Now, because at each subsequent step the working involves a linear cross-mean all entries in the f_{012i} column will have 1·68 as leading digits. Thus when calculating the column f_{012i} we do not carry or use the three digits 1·68 but work only with the digits to the right, as shown to the right of the dotted vertical line in the table.

The results given in the columns f_{012i} are the results of three different 4 point Lagrange interpolation polynomials, all of which involve the points (x_0, f_0); (x_1, f_1); (x_2, f_2) but use (x_3, f_3); (x_4, f_4); (x_5, f_5) respectively, in turn as the fourth point.

3.8.7 Advantages and suitable conditions for use of Aitken's method

If you work through the last example in § 3.8.6, checking the calculations and then compare it with the solutions given in § 3.7 it will become obvious that Aitken's method is quicker than Lagrange's method. Also Aitken's method shows automatically in the working, how the interpolates are converging and at the same time shows how the number of common significant figures is increasing at each stage. Further, if additional functional values are given these may be added to the table without need to do any recalculations, in other parts of the table.

In conclusion, when function values are given at unequal intervals, and not many interpolates are required from the given data, Aitken's method should be used.

3.9 EXAMPLES

1. The following values of x and corresponding function values are exact values of a cubic. Use them to obtain a third-order Lagrange interpolation polynomial. Use the polynomial to evaluate $f(1·5)$ exactly.

x	-1	0	2	5
$f(x)$	-2	2	20	332

2. Given the data

x	0	1	3	6
$f(x)$	-3	-5	18	200

form the cubic equation for a Lagrange interpolation polynomial.

3. From the following data

x	1·1	1·2	1·4	1·6	2·0
$f(x)$	1·4886	1·8681	3·1161	5·2281	13·2345

evaluate $f(1·235)$ to 4D using 5 point Lagrange interpolation.

4. Use Aitken's method for question No. 3.

5. Assuming $f(x)$ to be a function of the fourth degree find the value of $f(13)$ from the values

x	8	10	11	14	19
$f(x)$	4100	10 004	14 641	38 420	130 325

6. Using the data of question 5 find an expression for the Lagrange interpolation polynomial.

7. Use Lagrange 5 point interpolation formula to estimate $f(6)$ from the following table

x	0	3	5	9	12	
$f(x)$	0	252	446	1782	5976	(A.E.B.)

8. Using the following data

x	-3	-2	1	3	6
$f(x)$	-70	-19	2	26	317

obtain a value of $f(0)$. By consideration of the working and the interpolation polynomial, comment upon the accuracy of your result.

9. If $\quad x_0 = 0 \qquad\qquad x_1 = 1 \qquad\qquad x_2 = 4$

$\quad f(x_0) = -0.5001 \quad f(x_1) = -0.2731 \quad f(x_2) = -0.3841$

$\quad x_3 = 12$

$\quad f(x_3) = 0.4959$

use Lagrange interpolation to find $f(x)$ when $x = 10.79$, to 4D.

10. Using the data of question 9, use Aitken's method to evaluate $f(10.79)$ to 4D.

11. The four-point Lagrange interpolation formula is

$$y = L_{-1}y_{-1} + L_0 y_0 + L_1 y_1 + L_2 y_2 \text{ where}$$

$$L_0 = \frac{(x - x_{-1})(x - x_1)(x - x_2)}{(x_0 - x_{-1})(x_0 - x_1)(x_0 - x_2)} \text{ and } L_{-1}, L_1 \text{ and } L_2 \text{ are given by similar}$$

formulae.

(i) Given the table

x	-0.2	0.1	0.3	0.4
y	0.020 666	0.079 916	0.417 760	0.674 709

interpolate for y at $x = 0.2$.

(ii) If $x_0 - x_{-1} = x_1 - x_0 = x_2 - x_1 = h$ and $x = x_0 + sh$ show that $y = \frac{1}{16}[10(y_0 + y_1) - (y_{-1} + y_0 + y_1 + y_2)]$ at $s = \frac{1}{2}$. Given the table

x	5	6	7	8	9	10
y	51 140	61 387	71 667	81 983	92 337	102 733

interpolate for y at $x = 6.5$ and 7.5. (A.E.B.)

12. (i) Use Lagrange's interpolation formula to calculate $f(0)$ and $f(2.6)$ from the given table.

x	-2	-1	1	2	6
$f(x)$	-4	14	-4	-16	196

(ii) Given that all the values in the table are exact and that the function is a cubic, obtain the Lagrange interpolation polynomial.

13 From *I.A.T.* (page 59, C.7) write down the interpolation polynomial $P_2(x)$ and from it derive the form

$$\frac{(x - x_a)f(x_b) - (x - x_b)f(x_a)}{x_b - x_a}$$ where x_b and x_a are two values of x.

Defining this form as $f_{a,b}$ show that

$$P_3(x) = \frac{(x_2 - x)f_{01} - (x_0 - x)f_{12}}{x_2 - x_0} = f_{012}.$$

where f_{012} is the value at x of the Lagrangian interpolation polynomial which takes the values f_0, f_1, f_2 at the points x_0, x_1 and x_2 respectively. Hence, or otherwise, calculate the value of $f(x)$ when $x = 0$ from the given data, where the values of x are rounded and those of $f(x)$ are exact.

x	-1.9169	-0.6603	0.6966	2.1704	3.7777	
$f(x)$	1.09	1.11	1.13	1.15	1.17	(A.E.B.)

14. Obtain the coefficients A_0, A_2, A_4 and A_6 in the Lagrange four point interpolation formula

$$y = A_0(x - x_2)(x - x_4)(x - x_6) + A_2(x - x_0)(x - x_4)(x - x_6)$$
$$+ A_4(x - x_0)(x - x_2)(x - x_6) + A_6(x - x_0)(x - x_2)(x - x_4).$$

If the values of x have a common interval $2h$, show that

$$y_3 = \frac{1}{16}\{-y_0 + 9y_2 + 9y_4 - y_6\}.$$

Evaluate y for $x = 3, 5, 7$ from the following table and check your results by differencing at unit intervals.

x	0	2	4	6	8	10
y	0.000	-0.450	-0.698	-0.791	-0.768	-0.664

(A.E.B.)

4
Numerical Integration

4.1 INTRODUCTION

As we have seen in Volume I Numerical Integration is the evaluation of an integral of a function from a table of values of the function. Simpson's Rule and other similar formulae (See *I.A.T.*, page 70) can be derived geometrically, and express an approximate value of the integral in terms of simple multiples of equally spaced ordinates* (i.e. function values). These formulae are widely used and can give good accuracy in many cases. Unfortunately their geometrical derivation does not lead to any method of estimating the probable error in the result, and therefore, since it is essential in all serious numerical work to know the degree of accuracy which has been obtained, further investigation is required.

In this chapter we will apply finite difference methods to the problem of numerical integration and to the accuracy which can be obtained.

4.2 FINITE DIFFERENCE INTERPOLATION FORMULAE FOR ESTIMATING ACCURACY AND APPROXIMATE VALUE OF $\int f(x)dx$

Finite difference interpolation formulae are used to estimate values of a function at points in between those which are tabulated. This process, as we have shown, is also equivalent to the use of a polynomial which is approximately the same as the given function over a small range of values of the independent variable. In fact a finite difference formula is really an approximation to the equation of the curve which represents the function.

It follows that if we integrate this approximation to the function, then an approximate value of the integral of the function will be obtained. This is a more general approach to the problem and will be seen to provide a means of estimating not only the value of the integral, but also the accuracy which can be achieved. In other words we can determine how many of the significant figures in the answer are correct.

---

* This aspect of the subject has been dealt with in Vol. I, Chapter 9.

4.2.1 An important principle is demonstrated next.
    Before actually integrating any of the interpolation formulae it is important to understand that a change of independent variable from x to p is made which results in simplification of the work. Consider the following table:

x	$f(x)$	Δf	$\Delta^2 f$
0·850	0·138 333		
		11 545	
0·855	0·149 878		133
		11 678	
0·860	0·161 556		137
		12 815	
0·865	0·173 371		

In order to calculate values of x between 0·855 and 0·860 we set $x_0 = 0.855$ and let $x = x_0 + p.h$ where h is the interval of tabulation, i.e. $h = 0.005$, we thus have $x = 0.855 + 0.005\,p$, which is equivalent to $p = 200\,x - 171$. Gregory–Newton's forward difference interpolation formula now gives:

$$f_p = f_0 + p \cdot \Delta f_0 + \tfrac{1}{2}p(p - 1)\,\Delta^2 f_0$$

where $f_0 = 0.149\,878$, $\Delta f_0 = 0.011\,678$ and $\Delta^2 f_0 = 0.000\,137$ and the term in $\Delta^3 f_0$ has been omitted (it does not affect the sixth decimal place). This expression in terms of the variable p is now used as an approximation to the function $f(x)$. If we remember that $p = 200x - 171$ this approximation is equivalent to replacing $f(x)$ by:

$$f(x) \simeq f_0 + (200x - 171)\,\Delta f_0 + \tfrac{1}{2}(200x - 171)\,(200x - 172)\,\Delta^2 f_0$$

which is of the form $f(x) \simeq a_0 + a_1 x + a_2 x^2$. However the necessity of calculating a_0, a_1, a_2 may be avoided by using the approximation in its original form:

$$f_p = f_0 + p\Delta f_0 + \tfrac{1}{2}p(p - 1)\,\Delta^2 f_0$$

This expression is therefore to be regarded as being approximately the same as the function of x, but only within a limited range of values, and is expressed in terms of the variable p where p in this example is equal to $200x - 171$.

4.2.2 Suppose that in the same example we wish to evaluate the integral

$$I = \int_{0.855}^{0.865} f(x)\,\mathrm{d}x.$$

we must replace $f(x)$ by its equivalent f_p

giving $I = \int f_p \cdot \mathrm{d}x$

but $x = 0.855 + 0.005p$

Hence $\mathrm{d}x = 0.005 \cdot \mathrm{d}p$

and therefore $I = \int f_p \cdot \mathrm{d}p \times 0.005.$

The range of integration is from $x = 0.855$ to $x = 0.865$ which is the same as from $p = 0$ to $p = 2$, hence the integral can be evaluated as follows:

$$I = \int_{0.856}^{0.865} f(x) \, . \, \mathrm{d}x$$

$$= \int f_p \, \mathrm{d}x \qquad [\because f_p \equiv f(x)].$$

$$= 0.005 \int_0^2 f_p \, \mathrm{d}p \qquad [\because \mathrm{d}x = 0.005 \, . \, \mathrm{d}p]$$

$$= 0.005 \int_0^2 \{f_0 + p\Delta f_0 + \tfrac{1}{2}(p^2 - p)\Delta^2 f_0\} \, \mathrm{d}p$$

$$= 0.005 \, [f_0 p + \tfrac{1}{2}p^2 \Delta f_0 + \tfrac{1}{2}(\tfrac{1}{3}p^3 - \tfrac{1}{2}p^2) \, \Delta^2 f_0]_0^2$$

$$= 0.005 \, [2f_0 + 2\Delta f_0 + \tfrac{1}{3}\Delta^2 f_0]$$

$$= 0.005[0.299 \; 756 + 0.023 \; 356 + 0.000 \; 045 \; 7]$$

$$= 0.005 \times 0.323 \; 157 \; 7$$

$$= 0.001 \; 615 \; 79$$

keeping 6 significant figures.

The above result will be seen to be identical with that given by Simpson's Rule $\quad I = \tfrac{1}{3}h(y_0 + 4y_1 + y_2)$

$$\text{where} \quad y_0 = 0.149 \; 878$$

$$4y_1 = 0.646 \; 223$$

$$y_2 = 0.173 \; 371$$

$$\text{total} \qquad 0.969 \; 473$$

$$\therefore I = \tfrac{1}{3} \times 0.005 \times 0.969 \; 473$$

$$= 0.001 \; 615 \; 79 \qquad [\text{to 6S}]$$

The function used in the above example was $f(x) \equiv \tan x - 1$. The reader may care to integrate this expression to check that the above evaluations are correct to 6D.

In the next section we proceed to develop some of the standard integration formulae by integrating certain of the interpolation formulae which have been used in a previous chapter.

4.3 FORWARD DIFFERENCE INTEGRATION FORMULAE

The Gregory–Newton interpolation formula is

$$f_p = f_0 + p \, \Delta f_0 + \tfrac{1}{2} p \, (p - 1) \Delta^2 f_0 + \ldots + \binom{p}{n} \Delta^n f_0 + \ldots$$

where $f_0 = f(x_0)$; $f_p = f(x)$ and $x = x_0 + p \cdot h$ the function being tabulated at equal intervals h. In this formula we have the function $f(x)$ expressed in terms of the variable p which satisfies the linear relation $x = x_0 + ph$ in which x_0 and h are the constants. Differentiating this relation gives $dx/dp = h$ or $dx = h \cdot dp$. We can therefore evaluate the integral of $f(x)$ with respect to x for some range of values of x (usually a small number of complete intervals) by working in terms of the variable p as follows:

$$\int f(x)\, dx = \int f_p \cdot h \cdot dp \qquad (\because dx = h \cdot dp)$$

$$= h \cdot \int f_p dp. \qquad \text{(Since } h \text{ is a constant)}$$

and if the range of integration is from $x = x_0$ to $x = x_n$ which is the same as from $p = 0$ to $p = n$ we have:

$$\int_{x_0}^{x_n} f(x)\, dx = h \int_0^n f_p\, dp \ (h \text{ is a constant})$$

$$= h \int_0^n \left(f_0 + p\, \Delta f_0 + \frac{p^2 - p}{2} \Delta^2 f_0 + \frac{p^3 - 3p^2 + 2p}{6} \Delta^3 f_0 + \right.$$

$$\left. \frac{p^4 - 6p^3 + 11p^2 - 6p}{24} \Delta^4 f_0 + \dots \right) dp$$

[The Gregory–Newton formula]

$$= h \left[pf_0 + \tfrac{1}{2} p^2 \Delta f_0 + \left(\frac{p^3}{6} - \frac{p^2}{4} \right) \Delta^2 f_0 + \right.$$

$$\left. \left(\frac{p^4}{24} - \frac{p^3}{6} + \frac{p^2}{6} \right) \Delta^3 f_0 + \dots \right]_0^n$$

$$= h \left[nf_0 + \frac{n^2}{2} \Delta f_0 + \frac{n^2(2n - 3)}{12} \Delta^2 f_0 + \frac{n^2(p - 2)^2}{24} \Delta^3 f_0 + \dots \right] \qquad (1)$$

Putting $n = 1$ (i.e. using one interval or strip) we have

$$\int_{x_0}^{x_1} f(x) \cdot dx = h \left[f_0 + \tfrac{1}{2} \Delta f_0 - \tfrac{1}{12} \Delta^2 f_0 + \tfrac{1}{24} \Delta^3 f_0 + \dots \right] \qquad (2)$$

(*I.A.T.*, page 66 (Laplace))

Putting $n = 2$ (i.e. using two intervals or strips) we have

$$\int_{x_0}^{x_2} f(x)\, dx = h \left[2f_0 + 2 \Delta f_0 + \tfrac{1}{3} \Delta^2 f_0 - \tfrac{1}{90} \Delta^4 f_0 + \dots \right] \qquad (3)$$

In order to compare with previous results we can express formula (3) in terms of the ordinates. We have:

$$f_0 = y_0$$

$$\Delta f_0 = f_1 - f_0 = y_1 - y_0$$

$$\Delta^2 f_0 = \Delta f_1 - \Delta f_0 = y_2 - 2y_1 + y_0$$

D

hence (3) becomes

$$\int_{x_0}^{x_2} f(x) \, . \, dx = \frac{h}{3}[y_0 + 4y_1 + y_2 - \tfrac{1}{30}\Delta^4 f_0 + \ldots] \tag{4}$$

in which the coefficient of $\Delta^3 f_0$ is zero and the term in $\Delta^4 f_0$ has been left unchanged. If fourth- and higher-order differences can be ignored formula (4) is identical with Simpson's Rule. On the other hand, if we can only ignore fifth- and higher-order differences then the term $- (h/90) \Delta^4 f_0$ becomes an estimate of the error to be expected in using Simpson's Rule.

4.3.1 Worked example

Find $\int_0^{\pi/2} \sin x \, dx$ by tabulating $\sin x$ for $x = 0°(15°)120°$.

r	x_r	$\sin x$	Δ	Δ^2	Δ^3	Δ^4
0	0°	0·0				
			2588			
1	15°	0·2588		−176		
			2412		−165	
2	30°	0·5000		−341		24
			2071		−141	
3	45°	0·7071		−482		33
			1589		−108	
4	60°	0·8660		−590		40
			999		−68	
5	75°	0·9659		−658		44
			341		−24	
6	90°	1·0		−682		48
			−341		+24	
7	105°	0·9659		−658		
			−999			
8	120°	0·8660				

$h = (\pi/12) = 0·261\,80$ so the error involved in using Simpson's Rule to calculate the area of the first two strips is likely to be approximately $(h/90) \times \Delta^4 f_0$

$$= \frac{0·2618 \times 24}{90 \times 10^4} \simeq 0·07 \times 10^{-4}$$

This is smaller than we can expect the rounding errors to be in the tabulated values of the function.

Formula (3) may be used to evaluate $\int_0^{\pi/2} \sin x \, dx$ from the above difference table by applying it to three double strips in succession, giving:

$$\int_0^{\pi/2} \sin x \, dx \simeq h \left[2 \left(f_0 + f_2 + f_4 \right) + 2(\Delta f_0 + \Delta f_2 + \Delta f_4) \right.$$
$$\left. + \tfrac{1}{3} (\Delta^2 f_0 + \ldots) - \tfrac{1}{90} (\Delta^4 f_0 + \ldots) \right]$$

$f_0 = 0 \cdot 0$	$\Delta^2 f_0 = - \quad 176$
$f_2 = 0 \cdot 5000$	$\Delta^2 f_2 = - \quad 482$
$f_4 = 0 \cdot 8660$	$\Delta^2 f_4 = - \quad 658$
$1 \cdot 3660 \times 2 = 2 \cdot 7320$	$- 0 \cdot 1316 \times \tfrac{1}{3} = - 0 \cdot 0438 \; 7$

$\Delta f_0 = 0 \cdot 2588$	$\Delta^4 f_0 = \quad 24$
$\Delta f_2 = 0 \cdot 2071$	$\Delta^4 f_2 = \quad 40$
$\Delta f_4 = 0 \cdot 0999$	$\Delta^4 f_4 = \quad 48$
$0 \cdot 5658 \times 2 = 1 \cdot 1316$	$0 \cdot 0112 \times \tfrac{-1}{90} = - 0 \cdot 0001 \; 2$
$3 \cdot 8636$	$- 0 \cdot 0440$

Hence $\int_0^{\pi/2} \sin x \, . \, dx \simeq h \, (3 \cdot 8636 - 0 \cdot 0440)$

$$= 0 \cdot 261 \; 80 \times 3 \cdot 8196$$

$$= 0 \cdot 999 \; 97 \ldots$$

$$= 1 \cdot 0000 \, (\text{to 4D})$$

Retaining the fourth-order difference terms in the above calculation only affects the result by the amount $0 \cdot 261 \; 80 \times (- 0 \cdot 000 \; 12) = - 0 \cdot 000 \; 03$

Thus it appears that a proper use of an integrating formula can give results which are as accurate as the data from which they are calculated. In other words the process of integration tends to smooth out the effect of random errors in the data.

Many other formulae for numerical integration can be derived from the various interpolation formulae using forward, backward, or central differences. (See *I.A.T.*, pages 66–70, and Comrie's *Shorter Six Figure Tables*, pages 325–335.)

As a further illustration we will discuss two of the central difference formulae.

4.4   CENTRAL DIFFERENCE INTEGRATION FORMULAE AND THEIR USE

(i) Using Bessel's interpolation formula:

$$f_p = f_0 + p\delta f_{\frac{1}{2}} + B_2(\delta^2 f_0 + \delta^2 f_1) + B_3 \delta^3 f_{\frac{1}{2}} + B_4(\delta^4 f_0 + \delta^4 f_1) + \dots$$

where $\quad B_2 = \dfrac{p(p-1)}{4}$

$$B_3 = \dfrac{p(p-1)(p-\frac{1}{2})}{6}$$

$$B_4 = \dfrac{(p+1)(p)(p-1)(p-2)}{48}$$

$$\int_0^p B_2 \, dp = \left[\frac{p^3}{12} - \frac{p^2}{8}\right]_0^p = \frac{p^2(2p-3)}{24}$$

$$\int_0^p B_3 \, dp = \frac{1}{6}\left[\frac{p^4}{4} - \frac{p^3}{2} + \frac{p^2}{4}\right]_0^p = \frac{p^2(p-1)^2}{24}$$

$$\int_0^p B_4 \, dp = \frac{1}{48}\left[\frac{p^5}{5} - \frac{p^4}{2} - \frac{p^3}{3} + p^2\right] = \frac{p^2(6p^3 - 15p^2 - 10p + 30)}{30 \times 48}$$

$$\therefore \int_{x_0}^{x_1} f(x) \, . \, dx = h \int_0^1 (\text{above series}) \, . \, dp$$

$$= h\left[f_0 + \tfrac{1}{2}\delta f_{\frac{1}{2}} - \tfrac{1}{24}(\delta^2 f_0 + \delta f_1) + \frac{11}{1440}(\delta^4 f_0 + \delta^4 f_1) \dots\right]$$

$$= h\left[\frac{f_0 + f_1}{2} - \frac{\delta^2 f_0 + \delta^2 f_1}{24} + \frac{11(\delta^4 f_0 + \delta^4 f_1)}{1440} + \dots\right] \tag{5}$$

(See *I.A.T.*, page 66, formula 1)

(ii) Similarly by integrating Stirling's interpolation formula:

$$f_p = f_0 + \tfrac{1}{2}p(\delta f_{-\frac{1}{2}} + \delta f_{+\frac{1}{2}}) + \tfrac{1}{2}p^2\delta^2 f_0$$

$$+ \frac{p(p^2-1)}{12}(\delta^3 f_{-\frac{1}{2}} + \delta^3 f_{\frac{1}{2}}) + \frac{p^2(p^2-1)}{24}\delta^4 f_0 + \dots$$

it can be shown that:

$$\int_{x_{-1}}^{x_1} f(x) \, . \, dx = 2h(f_0 + \tfrac{1}{6}\delta^2 f_0 - \tfrac{1}{180}\delta^4 f_0 + \dots) \tag{6}$$

(See *I.A.T.*, page 66, formula 7)

4.4.1 Worked example 1

From the following table calculate:

(a) $\int_0^1 f(x).dx$, (b) $\int_0^2 f(x)\,dx$, (c) $\int_2^3 f(x)\,dx$, (d) $\int_2^4 f(x).dx$ and (e) $\int_0^{10} f(x)\,dx$

x	$f(x)$	Δ	Δ^2	Δ^3	Δ^4
0	1·0000				
		−486			
1	0·9514		+154		
		−332		−29	
2	0·9182		125		9
		−207		−20	
3	0·8975		105		6
		−102		−14	
4	0·8873		91		7
		−11		−7	
5	0·8862		84		1
		+73		−6	
6	0·8935		78		5
		151		−1	
7	0·9086		77		0
		228		−1	
8	0·9314		76		3
		304		+2	
9	0·9618		78		
		382			
10	1·0000				

(a) $\int_0^1 f(x).dx$ A forward difference formula must be used.

$$\int_0^1 f(x).dx = h(f_0 + \tfrac{1}{2}\Delta f_0 - \tfrac{1}{12}\Delta^2 f_0 + \tfrac{1}{24}\Delta^3 f_0 - \ldots)$$

$$= 1.(1\cdot 0000 - 0\cdot 0243 - 0\cdot 001\,28 - 0\cdot 000\,12) = 0\cdot 974\,30$$

(b) $\int_1^2 f(x)\,dx = h(f_1 + \tfrac{1}{2}\Delta f_1 - \tfrac{1}{12}\Delta^2 f_1 + \tfrac{1}{24}\Delta^3 f_1 - \ldots)$

$$= 1.(0\cdot 9514 - 0\cdot 0166 - 0\cdot 001\,04 - 0\cdot 000\,08) = 0\cdot 933\,72$$

Hence $\int_0^2 f(x)\,dx = \int_0^1 f(x)\,dx + \int_1^2 f(x).dx$

$$= 0\cdot 974\,30 + 0\cdot 933\,72$$

$$= 1\cdot 908\,0$$

(b) Alternatively, using (3)

$$\int_0^2 f(x)\,dx = h(2f_0 + 2\Delta f_0 + \tfrac{1}{3}\Delta^2 f_0 - \tfrac{1}{90}\Delta^4 f_0 + \ldots)$$

$$= 1 . (2{\cdot}0000 - 0{\cdot}0972 + 0{\cdot}005\,13 - 0{\cdot}000\,01 \ldots) = 1{\cdot}9079$$

(c) $\int_2^3 f(x)\,dx$ Central differences can be used.

$$\int_2^3 f(x)\,dx = h\left[\frac{f_2 + f_3}{2} - \frac{\delta^2 f_2 + \delta^2 f_3}{24} + \frac{11(\delta^4 f_2 + \delta^4 f_3)}{1440} - \cdots\right]$$

(using (5))

$$= 1 . \left[\frac{1{\cdot}8157}{2} - \frac{0{\cdot}0230}{24} + \frac{11 \times 0{\cdot}0015}{1440} - \cdots\right]$$

$$= 0{\cdot}9069$$

(d) $$\int_3^4 f(x)\,dx = h\left[\frac{f_3 + f_4}{2} - \frac{\delta^2 f_3 + \delta^2 f_4}{24} + \cdots\right]$$

$$= 1 . \left[\frac{1{\cdot}7848}{2} - \frac{0{\cdot}0196}{24} + \frac{11 \times 0{\cdot}0013}{1440} \cdots\right]$$

$$= 0{\cdot}8916$$

$$\therefore \int_2^4 f(x)\,dx = \int_2^3 f(x)\,dx + \int_3^4 f(x)\,dx = 1{\cdot}7985$$

(d) Alternative, using (6)

$$\int_2^4 f(x)\,dx = 2h\,(f_3 + \tfrac{1}{6}\delta^2 f_3 - \tfrac{1}{180}\delta^4 f_3 + \ldots)$$

$$= 2 . \left(0{\cdot}8975 + \frac{0{\cdot}0105}{6} - 0 + \ldots\right)$$

$$= 1{\cdot}7985$$

(e) $\int_0^2 f(x)\,dx = 1{\cdot}9079$ from (b) above.

$\int_2^3 f(x)\,dx = 0{\cdot}9069$ from (c) above.

$$\int_3^9 f(x)\,dx = 2h[(f_4 + f_6 + f_8) + \tfrac{1}{6}(\delta^2 f_4 + \delta^2 f_6 + \delta^2 f_8)$$

$$- \tfrac{1}{180}(\delta^4 f_4 + \delta^4 f_6 + \delta^4 f_8) + \ldots]$$

(using (6) from Stirling over 3 double strips and adding)

$$= 2\left[2{\cdot}7122 + \frac{0{\cdot}0245}{6} - \frac{0{\cdot}0016}{180} + \cdots\right]$$

$$= 5{\cdot}4325$$

$$\int_9^{10} f(x)\,dx = h[f_{10} - \tfrac{1}{2}\nabla f_{10} - \tfrac{1}{12}\nabla^2 f_{10} - \tfrac{1}{24}\nabla^3 f_{10} - \ldots]$$

(see I.A.T., page 66, formula 13)

$$= 1.\,[1 - 0{\cdot}0191 - 0{\cdot}0007 - 0\ldots]$$

$$= 0{\cdot}9803$$

$$\therefore \quad \int_0^{10} f(x)\,dx = \int_0^2 + \int_2^3 + \int_3^9 + \int_9^{10} f(x)\,dx$$

$$= 1{\cdot}9079 + 0{\cdot}9069 + 5{\cdot}4325 + 0{\cdot}9803$$

$$= 9{\cdot}2276$$

Note

Since the fourth differences are so small Simpson's Rule would be simpler to use, and gives results identical to (b), (d) and (e) above.

4.4.2   Worked example 2

$$\int_0^{\pi} \frac{\sin x}{x}\,.\,dx$$

Here the function to be integrated is known and can be tabulated outside the given range of integration and a central difference formula used, e.g. (6) from Stirling and in *I.A.T.*, page 66.

Summary of difference table using $h = \pi/6 = 0{\cdot}523\,5988$:

x	$\sin x/x$	δ^2	δ^4	δ^6	δ^8
0	1·000 000	−90 140	14 546	−2776	
$\pi/6$	0·954 930	−82 867	13 158	−2516	586
$\pi/3$	0·826 993	−62 436	9 254	−1670	63
$\pi/2$	0·636 620	−32 751	3 680	− 561	375
$2\pi/3$	0·413 496	+ 614	− 2 455	+ 723	−248
$5\pi/6$	0·190 986	31 524	− 7 867	1759	−365
π	0.	54 567	−11 520	2430	

$$\int\limits_{-1}^{+1} f_p \, \mathrm{d}p = 2 \left(1 + \frac{\delta^2}{6} - \frac{\delta^4}{180} + \ldots \right) f_0 \quad (I.A.T., \text{ page 66})$$

This formula is first applied to the range; $0 \leqslant x \leqslant \pi/3$ with

$$\left. \begin{array}{l} x_{-1} = 0 \ (\text{at } p = -1) \\[2mm] x_0 \ = \dfrac{\pi}{6} \ (\text{at } p = 0) \\[2mm] x_1 \ = \dfrac{\pi}{3} \ (\text{at } p = +1) \end{array} \right\} \quad \text{i.e. using } h = \pi/6$$

$\therefore$ Since $\quad \int f(x) \, \mathrm{d}x = h \cdot \int f_p \, \mathrm{d}p \quad$ (see §4.3)

Thus $\quad \displaystyle\int\limits_{0}^{\pi/3} \frac{\sin x \cdot \mathrm{d}x}{x} = h \int\limits_{-1}^{+1} f_p \, \mathrm{d}p$

$$= 2h \left(1 + \frac{\delta^2}{6} - \frac{\delta^4}{180} \ldots \right) f \left(\frac{\pi}{6} \right) \ldots \text{(a)}$$

Applying the same procedure to the next range of values $\pi/3 \leqslant x \leqslant 2\pi/3$ we would have:

$$x_{-1} = \frac{\pi}{3} \ (\text{at } p = -1)$$

$$x_0 = \frac{\pi}{2} \ (\text{at } p = 0)$$

$$x_1 = \frac{2\pi}{3} \ (\text{at } p = 1)$$

$$\therefore \quad \int\limits_{\pi/3}^{2\pi/3} \frac{\sin x \cdot \mathrm{d}x}{x} = 2h \left(1 + \frac{\delta^2}{6} - \frac{\delta^4}{180} \ldots \right) f \left(\frac{\pi}{2} \right) \ldots \qquad \text{(b)}$$

and similarly for the range $(2\pi/3) \leqslant x \leqslant \pi$ we have:

$$\int\limits_{2\pi/3}^{\pi} \frac{\sin x \cdot \mathrm{d}x}{x} = 2h \left(1 + \frac{\delta^2}{6} - \frac{\delta^4}{180} \ldots \right) f \left(\frac{5\pi}{6} \right) \ldots \qquad \text{(c)}$$

and adding the three results (a), (b) and (c) we finally obtain:

$$\int\limits_{0}^{\pi} \frac{\sin x \, \mathrm{d}x}{x} = 2h \left(1 + \frac{\delta^2}{6} - \frac{\delta^4}{180} + \frac{\delta^6}{1512} - \frac{23 \cdot \delta^8}{226\,800} + \ldots \right)$$

$$\left[f \left(\frac{\pi}{6} \right) + f \left(\frac{\pi}{2} \right) + f \left(\frac{5\pi}{6} \right) \right]$$

where $h = \pi/6 = 0.523\,598\,8$.

$$f\left(\frac{\pi}{6}\right) \quad 0.954\,930$$

$$f\left(\frac{\pi}{2}\right) \quad 0.636\,620$$

$$f\left(\frac{5\pi}{6}\right) \quad 0.190\,986$$

	$1.782\,536 \times 2$	$= +\,3.565\,072$
δ^2	$-82\,867$ $-32\,751$ $+31\,524$	
	$-84\,094 \div 3$	$= -\,0.028\,031\,3$
δ^4	$+13\,158$ $+\,3\,680$ $-\,7\,867$	
	$+\,8\,971 \div (-\,90)$	$= -\,0.000\,099\,7$
δ^6	$-\,2\,516$ $-\;\;\,561$ $+\,1\,759$	
	$-\,1\,318 \div 756$	$= -\,0.000\,001\,7$
δ^8	$+\;\;586$ $+\;\;375$ $-\;\;365$	
	$+\;\;596 \div (-\,5000) \simeq$	$-\,0.000\,000\,1$
		$3.536\,939\,2$

Thus:

$$\int_0^\pi \frac{\sin x\,\mathrm{d}x}{x} \simeq 0.523\,598\,8 \times 3.536\,939,\,2$$

$$\simeq 1.851\,937 \text{ (to 6 D).}$$

4.5   INDEFINITE INTEGRALS

We have, so far, discussed methods of estimating a complete area, i.e. the evaluation of a definite integral taken over a given range. But just in the same way as we think of any function of x changing continuously as the value of x increases, so we may consider the integral of $f(x)$ to be another function of x which varies as x increases. This section deals with the problem of evaluating numerically the indefinite integral of a tabulated function.

4.5.1.   The integral curve

In figure 4.5.1(a), OA $= a$ (an arbitrarily chosen constant) and OM $= x$ (variable) $\therefore$ the ordinate AB $= f(a)$ and MP $= f(x)$.

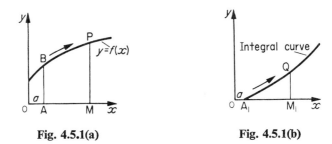

Fig. 4.5.1(a)   Fig. 4.5.1(b)

In figure 4.5.1(b) $OA_1 = a$ and $OM_1 = x$ and the ordinate M_1Q represents the area BAMP for each value of x. Then as x varies and P moves from B along the curve BP [i.e. $y = f(x)$], Q will move along a curve A_1Q which is called the integral curve of $y = f(x)$. If $\int f(x) \, dx = F(x) + C$, i.e. if $(d/dx) (F(x)) = f(x)$ then the equation of the integral curve A_1Q is $y = F(x) - F(a)$. The integral curve obtained will depend on the value chosen for 'a'. A different value of 'a' would give a new integral curve, but this would be the same as the curve A_1Q moved a distance parallel to OY, i.e. a new value of the arbitrary constant would be used. There are geometrical constructions by which the integral curve may be drawn (see Lipka: *Graphical and Mechanical Computation*, Wiley), and one integrator unit of an analogue computer gives the integral function as its output (see Vol. I § 1.1.6).

4.5.2   Numerical evaluation of the integral function

In order to tabulate values of the integral function at each of the given values of x, we must evaluate each of the integrals

$$A_1 = \int_{x_0}^{x_1} f(x) \, . \, dx, \, A_2 = \int_{x_1}^{x_2} f(x) \, . \, dx \ldots A_n = \int_{x_{n-1}}^{x_n} f(x) \, dx$$

and then form the cumulative totals, $A_1 + A_2$, $A_1 + A_2 + A_3, \ldots$, etc. If I_r is the value of the integral function at $x = x_r$ then

$$I_r = \int_{x_0}^{x_r} f(x) \, . \, dx + I_0$$

$$= \int_{x_0}^{x_1} f(x) \, dx + \int_{x_1}^{x_2} f(x) \, dx + \ldots + \int_{x_{r-1}}^{x_r} f(x) \, dx + I_0$$

$$= A_1 + A_2 + A_3 + A_4 + \ldots + A_r + I_0.$$

I_0, which is the value of the integral function at $x = x_0$, can usually be assumed to be zero, but in some cases a value may have to be assigned to it. It is the arbitrary constant which must be considered in connection with any indefinite integral.

Worked Example

Evaluate $\displaystyle\int_{0}^{x} \frac{\sin x \, . \, dx}{x}$ for $x = 0 \left(\dfrac{\pi}{6}\right) \pi$

The integral $\int_{0}^{\pi} (\sin x \, dx / x)$ has been evaluated in §4.4.2. The same table of differences can be used, but we must now evaluate the integral over each separate part of the range $0 < x < \pi$ at intervals of $\pi/6$.

Formula (5), § 4.4, is suitable for this purpose since it converges rapidly and uses only the even-order differences already tabulated.

$$\int_{0}^{x_1} f(x) \, dx = h \left\{ \frac{f_0 + f_1}{2} - \frac{\delta^2 f_0 + \delta^2 f_1}{24} + \frac{11 \, (\delta^4 f_0 + \delta^4 f_1)}{1440} \ldots \right\} \qquad (5)$$

The same formula is given in *I.A T.*, page 66, formula 1, in the more compact form:

$$\int_{0}^{1} f_p \, dp = \mu f_{\frac{1}{2}} - \frac{1}{12} \mu \delta^2 f_{\frac{1}{2}} + \frac{11}{720} \mu \delta^4 f_{\frac{1}{2}} - \frac{191}{604\,80} \mu \delta^6 f_{\frac{1}{2}} + \ldots$$

where μ is the 'Averaging' operator listed in *I.A.T.*, page 54 ,and its meaning is such that $\mu f_p = \frac{1}{2}(f_{p+\frac{1}{2}} + f_{p-\frac{1}{2}})$, i.e. $\mu f_{\frac{1}{2}} = \frac{1}{2}(f_0 + f_1)$, $\mu \delta^2 f_{\frac{1}{2}} = \frac{1}{2}(\delta^2 f_0 + \delta^2 f_1)$, etc.

This formula is applied to calculate the value of the integral over each separate tabulated interval, and the calculation may conveniently be arranged as follows:

x	$\mu f_{\frac{1}{2}}$	$-\dfrac{\mu\delta^2 f_{\frac{1}{2}}}{12}$	$+\dfrac{11\,\mu\delta^4}{720}$	$-\dfrac{191\,\mu\delta^6}{60,480}$	Totals	$\displaystyle\int_0^x f(x)\,dx$
0	—	—	—	—	—	0·0
$\dfrac{\pi}{6}$	0·977 465 0	+ 7 208 6	+ 211 6	+ 8 4	0·984 893 6	0·515 689 1
$\dfrac{\pi}{3}$	0·890 961 5	+ 6 054 3	+ 171 2	+ 6 6	0·897 193 6	0·985 458 6
$\dfrac{\pi}{2}$	0·731 806 5	+ 3 966 1	+ 98 8	+ 3 6	0·735 875 0	1·370 761 9
$\dfrac{2\pi}{3}$	0·525 058 0	+ 1 339 0	+ 9 4	− 0 3	0·526 406 1	1·646 387 5
$\dfrac{5\pi}{6}$	0·302 241 0	− 1 339 0	− 78 8	− 3 9	0·300 819 3	1·803 896 1
π	0·095 493 0	− 3 587 1	− 148 1	− 6 6	0·091 751 2	1·851 936 9
Totals	3·523 025 0	+ 13 641 9	+ 264 1	+ 7 8	3·536 938 8	—

The totals in the sixth column are multiplied by $h(\,=\pi/6=0\cdot523\,598\,8)$ and added to form the cumulative totals in column seven, which should now be rounded off to 6D. The entries in the last row can be used as an overall check on totals.

An independent check using a different integration formula is also advised. In this case we can use the previous example (§ 4.4.2) which agrees with the total of column six and the entry at $x = \pi$ in column seven. The above results at $x = \pi/3$ and $x = 2\pi/3$ can also be checked from the intermediate stages in § 4.4.2.

More systematic checking can be provided, and the reader is advised to refer to a more advanced text on Numerical Analysis, e.g. K. A. Redish, *An Introduction to Computational Methods*, E.U.P., Example 43, pages 102–103.

4.6 EXAMPLES

1. (a) Difference the given table of values of $f(x)$, and hence estimate the error involved in using Simpson's Rule to evaluate $\int_1^6 f(x) \, . \, dx$.

x	$f(x)$	x	$f(x)$
1	0·909 09	4	1·538 46
1·5	1·224 49	4.5	1·487 60
2	1·428 57	5	1·428 57
2·5	1·538 46	5·5	1·366 46
3	1·578 95	6	1·304 35
3·5	1·573 03		

(b) Evaluate the given integral as accurately as possible.

2. (a) If $y = 1 + \log_e x$ show that $(d^4 y/dx^4) = -(2/27)$ when $x = 3$.

(b) Hence show that if y is tabulated to 6D, giving fourth differences of about 100. the interval of tabulation must be approximately 0·2 for values of x near 3.

(c) What would you expect the values of the sixth differences to be? (apart from rounding errors).

See I.A.T. p. 63. § 7 $\delta^4 f_0 \simeq h^4 . \, d^4 f/dx^4$ for small h.

3. (a) Tabulate values of $1 + \log_e x$ for $x = 2\cdot6(0\cdot2)4\cdot0$ and form the difference table.

(b) Evaluate $\int_3^x (1 + \log_e x) \, dx$ for $x = 3\cdot0(0\cdot1)3\cdot6$.

(c) Check your answers by tabulating values of $x \log_e x$.

(Note. $\int(1 + \log_e x) \, dx = x \log_e x$)

4. The curve $y = f(x)$ is rotated about the axis OX. Find the volume generated and the position of its C.G. by tabulating values of y^2 and xy^2 from the given table of values.

x	$f(x)$
0	1·0
0·1	1·004 837
0·2	1·018 731
0·3	1·040 818
0·4	1·070 320
0·5	1·106 531
0·6	1·248 812

Tabulate y^2 and xy^2 to 5D and evaluate $V = \pi \int_0^{0\cdot6} y^2 \, dx$, and $\bar{x}$, given that $V \times \bar{x} = \pi \int_0^{0\cdot6} xy^2 \, dx$.

5. Find the value of $E = \int_0^{\pi/2} \sqrt{(1 - \frac{1}{4}\sin^2 x)}\,.\,dx$ correct to 4D by tabulating values of $\sqrt{(1 - \frac{1}{4}\sin^2 x)}$ for $0 \leqslant x \leqslant \pi/2$, choosing a suitable interval for tabulation.

6. Tabulate $\sqrt[4]{(1 + x^2)}$ for $x = 0(0.25)2$ to 4D. Evaluate $\int_0^2 \sqrt[4]{(1 + x^2)}\,.\,dx$ and comment on the accuracy of your answer.

(Note the symmetry of the function about $x = 0$).

7. By dividing the range $(0.1, 0.5)$ into four equal parts evaluate $\int_{0.1}^{0.5} e^{-x^2}\,dx$ to 4D.

(For values of e^{-x^2} for which the tables are inadequate use

$$e^{-x^2} = 1 - x^2 + \frac{x^4}{2!} - \frac{x^6}{3!} + \frac{x^8}{4!}\dots).$$ (Cambridge 1965)

8. Using the following table which lists $f(0.5)$ and $f(0.6)$ and the significant odd order differences, evaluate $\int_{0.2}^{0.6} f(x)\,dx$.

x	$f(x)$	$\delta^2_{\frac{1}{2}}$	$\delta^3_{\frac{1}{2}}$	$\delta^5_{\frac{1}{2}}$	$\delta^7_{\frac{1}{2}}$
		190	-100	10	-10
0.5	1763.57				
		2413	-102	108	-12
0.6	1787.70				
		4534	4	194	-11

(Cambridge 1966)

9. Tabulate $(x - 2)\sqrt{(4 + x^3)}$ for $0.8\,(-0.1) - 0.4$. Difference the table and use it to assist in the evaluation of $\int_0^{0.5} (x - 2)\sqrt{(4 + x^3)}\,.\,dx$ correct to 5D.   (A.E.B. 1965)

10. Tabulate to 6D, $y = e^{-x}/(2 + x)$ for $0.10(0.05)0.50$. Difference your table and hence evaluate $\int_{0.2}^{0.4} e^{-x}\,dx/(2 + x)$.   (A.E.B. 1966)

11. By integrating the Bessel interpolation formula show that

$$(1/h) \int_{x_0}^{x_1} f\,dx = (\mu - \tfrac{1}{12}\mu\delta^2 + \tfrac{11}{720}\mu\delta^4 - \dots)f_{\frac{1}{2}}.$$

Given that

$\int_0^{0\cdot4} y\, dx = 0\cdot4906$ use the following table to evaluate

$\int_0^{0\cdot6} y\, dx$ correct to 3D.

x	0·2	0·3	0·4	0·5	0·6	0·7	0·8
y	1·3117	1·4843	1·6075	1·7033	1·7818	1·8482	1·9058

(A.E.B. 1964)

12. Evaluate $\int_0^{0\cdot6} \sqrt{(x^2 + x + 7)}\, dx$ to 7D. Comment on the accuracy of your seventh decimal place. (A.E.B. 1963)

13. The function $(\sin x/x)$ is tabulated with its second differences. Evaluate $\int_{0\cdot6}^{0\cdot7} (\sin x/x)\, dx$ correct to 7D.

x	$\sin x/x$	δ^2
0·4	0·973 545 8	−31 731
0·5	0·958 851 1	−30 856
0·6	0·941 070 8	−29 795
0·7	0·920 311 0	−28 561
0·8	0·896 695 1	−27 160
0·9	0·870 363 2	−25 603

(A.E.B. 1962)

14. Tabulate $y = (\sqrt{x}) \cos x$. Difference your table and hence evaluate $\int_{0\cdot5}^{0\cdot6} (\sqrt{x}) \cos x\, dx$, correct to 5D. (A.E.B. 1960)

5
Numerical Differentiation

5.1 INTRODUCTION

In dealing with a short arc of a curve near any particular point P of the curve, we must use an approximation which is appropriate to our purpose.

As a first approximation, the curve very close to P may be regarded as a straight line for some purposes. This is used in elementary calculus in discussing the gradient of the curve at P and to evaluate the length of an arc.

Again we sometimes assume that the curve very close to P approximates to an arc of a circle. This method is used to find the radius of curvature at P.

In this chapter our aim is to obtain a number of formulae for the calculation of derivatives of a function from a given set of numerical values of the function. Since this is essentially the same problem as that of calculating the gradient at a point on the curve which represents the function, we will begin with a fairly obvious geometrical approach which enables the simpler formulae to be obtained easily. This will be followed by our normal approach via interpolating polynomials which will be used as approximations to portions of the curve. Differentiation of these approximate equations leads to the required formulae for calculation of derivatives.

5.1.1 Assuming that the curve is very nearly parabolic, approximations can be obtained which are useful to consider first. Assume that its equation is of the form:

$$y = a_0 + a_1 x + a_2 x^2.$$

Referring to Fig. 5.1(a) and assuming that the arc is a parabola, we have three equally spaced ordinates cutting the curve in P, Q and R. In this case it can be proved that the tangent at Q is parallel to the chord PR. That is the gradient of the curve at Q is equal to the gradient of the chord PR.

This result remains approximately true, in general, when PQR is a small arc of any curve. The accuracy of such an approximation depends on the choice of a suitable value of h, the distance between ordinates, or the interval of tabulation. Care must also be taken near any special points, particularly where Q is at or near a point of inflexion.

Practical use is made of this property of curves by draughtsmen when a tangent has to be constructed as accurately as possible at a given point on a curve. We can use it to find an approximate value of the gradient, at certain points, in terms of the first differences in the tabulated values:

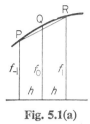

Fig. 5.1(a)

(a) The gradient at a tabulated point.
   In this case the ordinates to P, Q and R represent f_{-1}, f_0 and f_1.
   The gradient at Q $\simeq$ the gradient of the chord PR

$$\therefore \qquad f_0' \simeq \frac{f_1 - f_{-1}}{2h}$$

i.e.
$$f_0' \simeq \frac{1}{2h}(\delta f_{\frac{1}{2}} + \delta f_{-\frac{1}{2}}) \text{ or } \frac{1}{h}\mu \, \delta f_0$$

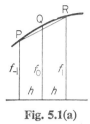

Fig. 5.1(b)

(b) The gradient at a "half-way" point.
   In this case ordinates to P and R represent two consecutive tabulated values f_0 and f_1 and the ordinate to Q is the mid-ordinate $f_{\frac{1}{2}}$ as shown in Fig. 5.1(b).
   The gradient at Q $\simeq$ the gradient of the chord PR

$$\therefore \qquad f_{\frac{1}{2}}' \simeq \frac{f_1 - f_0}{h}$$

i.e.
$$f_{\frac{1}{2}}' \simeq \frac{1}{h} \cdot \delta f_{\frac{1}{2}}$$

5.1.2 There are two main sources of error in using the above very simple formulae, and in using all methods of approximation, which it is important to remember. There are errors in the data, and errors due to the particular formula which has been selected. Errors in the data may become especially important in differentiation formulae when h the interval of tabulation is small.

Worked example

Consider the following example in which various approximations to $f'(2)$ are obtained where $f(x) \equiv 1/x$.

	x	$f(x)$	δf	h	$f'(2)$	possible rounding off error
(a)	1·8	0·5556	−0·1011	0·4	−0·253	±0·000 25
	2·2	0·4545				
(b)	1·9	0·5263	−0·0501	0·2	−0·2505	±0·000 5
	2·1	0·4762				
(c)	1·95	0·5128	−0·0250	0·1	−0·250	±0·001
	2·05	0·4878				
(d)	1·995	0·5013	−0·0025	0·01	−0·25	±0·01
	2·005	0·4988				

$f(x)$ is rounded to 4D, therefore the error in δf may possibly be as much as ± 1 in the last figure and this when multiplied by $1/h$ could cause the errors listed in the final column.

The answer in (a) is wrong in the third significant figure. This is not due to rounding errors, but is caused by using too big a value of h for this simple formula (formulae used later in the Chapter would give greater accuracy).

The error in (b) comes in the fourth significant figure and is due partly to rounding and partly to the formula.

The answer in (c) is correct to three significant figures but rounding errors could have effected the third figure.

The answer in (d) is correct to only two significant figures and rounding errors could even have affected the second figure.

It should be clear that as the value of h decreases, the magnitude of the first differences also decreases. Therefore, when a small value of h is used the number of significant figures in each first difference is less than the number of significant figures in the data. The calculated value of the gradient is subject to the same loss of significant figures since it is obtained from the first difference. This problem of accuracy will be referred to again in § 5.5.

5.2 EXPRESSIONS FOR ERROR ESTIMATES USING THE CALCULUS

An estimate of the error involved in the above simple formulae can be obtained by expressing the same ideas analytically in the notation of calculus. The three ordinates to P, Q, R in Figure 5.1(a) are equal to

$$f(x - h), f(x), f(x + h)$$

respectively, and the gradient of the chord PR is therefore $\{f(x+h)-f(x-h)\}/2h$ and we have:

(a) where $y = f(x)$ is the actual function in the problem as $h \to 0$ the gradient of the chord approaches the gradient of the tangent at Q

i.e. $$f'(x) = \lim_{h \to 0} \frac{f(x+h)-f(x-h)}{2h}$$

(b) where $y = f(x)$ is an approximation to the function in the problem

$$f'(x) \simeq \frac{f(x+h)-f(x-h)}{2h}$$

provided a suitably small value of h is used.

(c) Using Taylor's theorem to expand $f(x+h)$ and $f(x-h)$ we have:

$$f(x+h) = f(x) + hf'(x) + \tfrac{1}{2}h^2 f''(x) + \tfrac{1}{6}h^3 f'''(x) + \ldots$$

$$f(x-h) = f(x) - hf'(x) + \tfrac{1}{2}h^2 f''(x) - \tfrac{1}{6}h^3 f'''(x) + \ldots$$

$$\therefore f(x+h) - f(x-h) = 2h \{f'(x) + \tfrac{1}{6}h^2 f'''(x) + \ldots\}$$

$$\therefore f'(x) \simeq \frac{f(x+h)-f(x-h)}{2h} - \tfrac{1}{6}h^2 f'''(x) - \ldots$$

provided that the neglected terms in h^4 and higher powers of h are small enough.

Thus the error involved in using the formula of 5.1(a) is approximately equal to $\tfrac{1}{6}h^2 f'''(x)$ which is of the same order of magnitude as $(1/6h)\,\delta^3 f(x)$. (See *I.A.T.* page 63 § 7).

This result will appear again later, (§ 5.3.2) in the second term of formula (4).

5.3   CENTRAL DIFFERENCE FORMULA FOR DIFFERENTIATION

Having seen some of the difficulties associated with numerical differentiation and investigated methods of obtaining simple formulae from geometrical considerations we now continue by looking into the results which can be obtained from the differentiation of the various interpolating polynomials.

In applying tabulated differences to differentiation (i.e. calculation of gradients) it is important to use central differences whenever possible. They make use of the data available on both sides of the point under discussion and tend to give the greatest possible accuracy. In addition it will be noticed that the central difference formulae most frequently used contain fewer terms and converge more quickly than the corresponding forward difference formulae.

Each interpolating polynomial f_p is a way of expressing the function $f(x)$ approximately in terms of the variable p where $x = x_0 + ph$. So $(\mathrm{d}p/\mathrm{d}x) = (1/h)$.

Hence when we differentiate $f(x)$ and f_p we must have:

$$\frac{d}{dx}\left[f(x)\right] = \frac{d}{dx}\left[f_p\right]$$

$$= \frac{d}{dp}\left[f_p\right] \times \frac{dp}{dx}$$

i.e. $$f'(x) = \frac{1}{h} \cdot \frac{df_p}{dp}$$

Here the prime (') indicates differentiation with respect to x and the symbol f'_p is used to indicate the value of $f'(x)$ when expressed in terms of p in the same way as f_p equals $f(x)$.

$$\therefore \qquad f'(x) = f'_p = \frac{1}{h} \cdot \frac{df_p}{dp}$$

5.3.1 Using Bessel's interpolation formula:

$$f(x) \equiv f_0 + p\delta f_{\frac{1}{2}} + \tfrac{1}{4}p(p-1)(\delta^2 f_0 + \delta^2 f_1) + \tfrac{1}{6}p(p-1)(p-\tfrac{1}{2})\delta^3 f_{\frac{1}{2}} +$$

$$\frac{(p+1)p(p-1)(p-2)}{48}(\delta^4 f_0 + \delta^4 f_1) + \ldots$$

and by differentiating the series with respect to p we get

$$f'(x) \equiv f'_p = \frac{1}{h}\left\{\delta f_{\frac{1}{2}} + \frac{2p-1}{4}(\delta^2 f_0 + \delta^2 f_1) + \frac{3p^2 - 3p + \frac{1}{2}}{6} \cdot \delta^3 f_{\frac{1}{2}} + \right.$$

$$\left. \frac{(2p-1)(p^2-p-1)}{24}(\delta^4 f_0 + \delta^4 f_1) + \ldots\right\} \qquad (1)$$

This formula can be used to calculate the gradient at any point between x_0 and x_1, but it is particularly convenient for use at the midway point $x = x_0 + \tfrac{1}{2}h$ because on substituting $p = \tfrac{1}{2}$ we have:

$$f'(x_0 + \tfrac{1}{2}h) \equiv f'_{\frac{1}{2}} = \frac{1}{h}\left\{\delta f_{\frac{1}{2}} - \frac{1}{24}\delta^3 f_{\frac{1}{2}} + \frac{3}{640}\delta^5 f_{\frac{1}{2}} - \ldots\right\} \qquad (2)$$

5.3.2 Similarly by differentiating Stirling's interpolation formula

$$f_p = f_0 + \tfrac{1}{2}p(\delta f_{-\frac{1}{2}} + \delta f_{\frac{1}{2}}) + \tfrac{1}{2}p^2\delta^2 f_0 + \frac{p(p^2-1)}{12}(\delta^3 f_{-\frac{1}{2}} + \delta^3 f_{\frac{1}{2}}) +$$

$$\frac{p^2(p^2-1)}{24}\delta^4 f_0 +$$

we can obtain:

$$f'(x) \equiv f'_p = \frac{1}{h} \left\{ \tfrac{1}{2}(\delta f_{-\frac{1}{2}} + \delta f_{\frac{1}{2}}) + p\delta^2 f_0 + \frac{3p^2 - 1}{12} (\delta^3 f_{-\frac{1}{2}} + \delta^3 f_{\frac{1}{2}}) + \right.$$
$$\left. \frac{p(2p^2 - 1)}{12} \delta^4 f_0 + \dots \right\} \quad (3)$$

This again can be used at intermediate points but is particularly suited for use at a tabular point because on substituting $p = 0$ we have:

$$f'(x_0) \equiv f'_0 = \frac{1}{h} \left\{ \tfrac{1}{2}(\delta f_{-\frac{1}{2}} + \delta f_{\frac{1}{2}}) - \frac{1}{12} (\delta^3 f_{-\frac{1}{2}} + \delta^3 f_{\frac{1}{2}}) + \right.$$
$$\left. \frac{1}{60} (\delta^5 f_{-\frac{1}{2}} + \delta^5 f_{\frac{1}{2}}) \dots \right\} \quad (4)$$

which can be written more compactly, using μ (the averaging operator) as in Chapter 4 § 4.5.2, as follows:

$$f'_0 = \frac{1}{h} \left\{ \mu \delta f_0 - \frac{1}{6} \mu \delta^3 f_0 + \frac{1}{30} \mu \delta^5 f_0 \dots \right\} \quad (4)$$

It will be noticed that the first term in this formula is identical with the formula in § 5.1.1(a) and that the second term agrees with that in § 5.2(c).

5.4   FORWARD AND BACKWARD DIFFERENCE FORMULAE

5.4.1 For a point near the beginning of the table, in cases where it is necessary to calculate the derivative, a forward difference formula will have to be used. On differentiating the Gregory–Newton formula

$$f_p = f_0 + p\Delta f_0 + \frac{p(p - 1)}{2} \Delta^2 f_0 + \frac{p(p - 1)(p - 2)}{6} \Delta^3 f_0 + \dots$$

we obtain:

$$f'(x) \equiv f'_p = \frac{1}{h} \left\{ \Delta f_0 + \frac{2p - 1}{2} \Delta^2 f_0 + \frac{3p^2 - 6p + 2}{6} \Delta^3 f_0 + \dots \right\} \quad (5)$$

for use at any value of p in the range $0 \leqslant p \leqslant 1$.

Putting $p = 0$ we have at the tabular point x_0

$$f'_0 = \frac{1}{h} \left\{ \Delta f_0 - \tfrac{1}{2}\Delta^2 f_0 + \tfrac{1}{3}\Delta^3 f_0 - \tfrac{1}{4}\Delta^4 f_0 + \tfrac{1}{5}\Delta^5 f_0 + \dots \right\} \quad (6)$$

Putting $p = 1$ we have at the next tabular point x_1

$$f'_1 = \frac{1}{h} \left\{ \Delta f_0 + \tfrac{1}{2}\Delta^2 f_0 - \tfrac{1}{6}\Delta^3 f_0 + \tfrac{1}{12}\Delta^4 f_0 - \tfrac{1}{20}\Delta^5 f_0 + \dots \right\} \quad (7)$$

and putting $p = \frac{1}{2}$ we have for the halfway point

$$f'_{\frac{1}{2}} = \frac{1}{h} \{ \Delta f_0 - \tfrac{1}{24} \Delta^3 f_0 + \tfrac{1}{24} \Delta^4 f_0 \ldots \} \tag{8}$$

5.4.2 For a point near the end of the table, in cases where it is necessary to calculate the derivative, a backward difference formula must be used. The following are obtained from the Gregory–Newton backward difference interpolation formula:

$$f'_0 = \frac{1}{h} (\nabla f_0 + \tfrac{1}{2}\nabla^2 f_0 + \tfrac{1}{3}\nabla^3 f_0 + \tfrac{1}{4}\nabla^4 f_0 + \tfrac{1}{5}\nabla^5 f_0 + \ldots) \tag{9}$$

$$f'_{-1} = \frac{1}{h} (\nabla f_0 - \tfrac{1}{2}\nabla^2 f_0 - \tfrac{1}{6}\nabla^3 f_0 - \tfrac{1}{12}\nabla^4 f_0 - \tfrac{1}{20}\nabla^5 f_0 - \ldots) \tag{10}$$

$$f'_{-\frac{1}{2}} = \frac{1}{h} (\nabla f_0 - \tfrac{1}{24}\nabla^3 f_0 - \tfrac{1}{24}\nabla^4 f_0 \ldots) \tag{11}$$

5.5 ROUNDING ERRORS

It should again be observed that all the formulae for numerical differentiation involve a factor $1/h$ multiplying a series containing differences of ascending order. Rounding errors in the data lead to quite large possible errors in the higher differences. (See Vol. I, Chapter 5, §5.1.8 and I.A.T., page 55 B.2).

This possible accumulation of rounding errors is most important when as in differentiation the series involved converge slowly and consequently it can be difficult to assess the degree of accuracy attained. There will almost always be some loss of significant figures because the first, and largest, term in all the formulae is $(1/h) \times \delta f$ (or $(1/h) \times \Delta f$) and the number of significant figures in the first difference column is in general less than the number of significant figures given in the data. Moreover, the smaller the interval of tabulation the greater will be this loss of significant figures.

In some cases it may therefore even be advisable to use a comparatively large value of h in order to achieve the desired accuracy.

This increase of relative error due to losing significant figures only occurs in differentiation. In the other applications of finite difference methods to interpolation and integration the first (and most significant), term in the calculation is based on actual tabulated values of the function. Thus the accuracy attainable in integration and interpolation is comparable with the accuracy of the given data, whereas in differentiation results depend mainly on the accuracy of the first difference columns.

Worked Example

Using the following table of values, calculate $f'(2)$ and $f'(2\cdot05)$.

x	$f(x)$	δf (± 1)	$\delta^2 f$ (± 2)	$\delta^3 f$ (± 4)	$\delta^4 f$ (± 8)	$\delta^5 f$
1·7	0·588 24					
		− 32 68				
1·8	0·555 56		344			
		− 29 24		− 52		
1·9	0·526 32		292		9	
		− 26 32		− 41		− 2
2·0	0·500 00		251		7	
		− 23 81		− 34		− 3
2·1	0·476 19		217		4	
		− 21 64		− 30		+ 5
2·2	0·454 55		187		9	
		− 19 77				
2·3	0·434 78		166			
		− 18 11				
2·4	0·416 67					

$f(x)$ has been taken to be $1/x$ so that accuracy of the answers can be readily checked. The rounding errors in $f(x)$ are not greater than $\frac{1}{2}$ unit in the fifth decimal place. The greatest possible errors in the differences due to this cause are indicated at the top of the columns, but are very unlikely to be so large in practice.

For $f'(2)$

(a) Using (4) from § 5.3.2, with $x_0 = 2$ and $h = 0\cdot1$ we have

$$f'(2) = f_0 = \frac{1}{h}\left\{\frac{\delta f_{-\frac{1}{2}} + \delta f_{\frac{1}{2}}}{2} - \frac{\delta^3 f_{-\frac{1}{2}} + \delta^3 f_{\frac{1}{2}}}{12} + \frac{\delta^5 f_{-\frac{1}{2}} + \delta^5 f_{\frac{1}{2}}}{60} - \cdots\right\}$$

$$= \frac{1}{0\cdot1}\left\{\frac{- 0\cdot050\ 13}{2} - \frac{- 0\cdot000\ 75}{12} + \text{negligible terms}\right\}$$

$$= 10\ \{- 0\cdot025\ 0\overset{\bullet}{6}5 + 0\cdot000\ 062\}$$

$$= - 0\cdot250\overset{\bullet}{0}\ \text{(to 4D)}$$

* This calculation appears to give us 4 correct significant figures but in an extreme case there could have been a rounding error of ± 1 in the fifth decimal place of the first term resulting in a similar error in the fourth decimal place of the final answer. The final contribution of errors due to rounding do not affect the first five decimal places of the second and third terms.

(b) Using (6) from § 5.4.1 we have, with $x_0 = 2$, $h = 0{\cdot}1$

$$f'(2) = f_0' = \frac{1}{h}\left\{ \Delta f_0 - \tfrac{1}{2}\Delta^2 f_0 + \tfrac{1}{3}\Delta^3 f_0 - \ldots \right\}$$

$$= \frac{1}{0{\cdot}1}\left\{ -0{\cdot}023\,81 - \frac{0{\cdot}002\,17}{2} + \frac{-0{\cdot}000\,30}{3} - \frac{0{\cdot}000\,09}{4} \ldots \right\}$$

$$= \frac{1}{0{\cdot}1}\left\{ -0{\cdot}023\,8\overset{\bullet}{1} - 0{\cdot}001\,0\overset{\bullet}{8}5 - 0{\cdot}000\,1\overset{\bullet}{0} - 0{\cdot}000\,0\overset{\bullet}{2}2 \ldots \right\}$$

$$= 10 \times (-0{\cdot}025\,0\overset{\bullet}{2})$$

$$= -0{\cdot}250 \text{ (to 3D)}$$

* In this example the errors due to rounding might affect each term by anything up to ± 1 in the fifth decimal place, and convergence is so slow that fifth (or higher) differences could also affect the same figure.

For $f'(2{\cdot}05)$

(c) Using (2) from § 5.3.1, with $x_0 = 2$, $h = 0{\cdot}1$ we have:

$$f'(2{\cdot}05) = f_{\frac{1}{2}}' = \frac{1}{h}\left\{ \delta f_{\frac{1}{2}} - \frac{1}{24}\,\delta^3 f_{\frac{1}{2}} + \frac{3}{640}\,\delta^5 f_{\frac{1}{2}} - \ldots \right\}$$

$$= \frac{1}{0{\cdot}1}\left\{ -0{\cdot}023\,81 - \frac{-0{\cdot}000\,34}{24} + \text{negligible terms} \right\}$$

$$= 10\left\{ -0{\cdot}023\,81 + 0{\cdot}000\,014 \right\}$$

$$= -0{\cdot}2380 \text{ (error} < 0{\cdot}0001)$$

(d) Using (8) from § 5.4.1

$$f'(2{\cdot}05) = f_{\frac{1}{2}}' = \frac{1}{h}\left\{ \Delta f_0 - \frac{1}{24}\Delta^3 f_0 + \frac{1}{24}\Delta^4 f_0 - \ldots \right\}$$

$$= \frac{1}{0{\cdot}1}\left\{ -0{\cdot}023\,81 - \frac{-0{\cdot}000\,30}{24} + \frac{0{\cdot}000\,09}{24} \ldots \right\}$$

$$= 10\left\{ -0{\cdot}023\,81 + 0{\cdot}000\,012 + 0{\cdot}000\,004 \ldots \right\}$$

$$= -0{\cdot}2379 \text{ (error} < 0{\cdot}0001)$$

$$\left(\text{N.B. } f'(2{\cdot}05) = -\frac{1}{(2{\cdot}05)^2} = -0{\cdot}237\,954\right)$$

In the above example attention has been drawn to the greatest possible effect of errors due to the rounding off of the given data. This maximum error can only occur if the errors in $f_0, f_1, f_2, \ldots$ etc. are alternately $+\frac{1}{2}, -\frac{1}{2}, +\frac{1}{2}, \ldots$, etc., in the last decimal place, and this is extremely unlikely to happen.

Rounding errors are usually 'random' errors, and statistical methods have been used to estimate the 'probable errors' which could result from them. Table D.2, page 62, *I.A.T.*, gives some guidance on accuracy to be expected in using central difference formulae for calculating derivatives, e.g. errors in calculating hf_0' using (4) and $hf_{\frac{1}{2}}'$ using (2) above are probably less than 0·34 (i.e. one-third) and 0·49 (i.e. one-half) respectively, in units of the last figure. Referring to the worked example, the probable error in (a) $f'(2)$ is $< 0.000\,03$ and the probable error in (c) $f'(2·05)$ is $< 0.000\,05$. Rounding to 4D is therefore justified in these cases.

5.6 REPEATED DIFFERENTIATION

Finite difference formulae which can be used to calculate values of the second derivative of a tabulated function are derived by differentiating a second time the standard interpolating formula. The method used is the same as in the previou: sections, and the central difference formula obtained from Stirling's formula iss

$$f_p'' = \frac{1}{h^2}\left\{ \delta^2 f_0 + p\mu\delta^3 f_0 + \frac{6p^2 - 1}{12}\delta^4 f_0 \ldots \right\} \tag{12}$$

which could be used at intermediate points, or for use at tabular points we have on putting $p = 0$:

$$f_0'' = \frac{1}{h^2}\left\{ \delta^2 f_0 - \frac{1}{12}\delta^4 f_0 + \frac{1}{90}\delta^6 f_0 - \ldots \right\} \tag{13}$$

Other formulae for first, second and higher derivatives are to be found in *I.A.T.*, pages 61–64.

Worked Example

From the following table of values calculate the values of $f'(x)$ for $x = 3·0(0·4)4·6$ and evaluate $f'(3·9)$ and $f''(4)$.

x	$f(x)$	δ	δ^2	δ^3	δ^4	δ^5
2·4	0·183 22					
		+ 158 42				
2·8	0·341 64		− 1682			
		141 60		443		190
3·2	0·483 24		− 1239		− 165	
		129 21		278		74
3·6	0·612 45		− 961		− 91	
		119 60		187		37
4·0	0·732 05		− 774		− 54	
		111 86		133		20
4·4	0·843 91		− 641		− 34	
		105 45		99		
4·8	0·949 36		− 542			
		100 03				
5·2	1·049 39					

Using $f'_{\frac{1}{2}} = \frac{1}{h}\left\{ \delta f_{\frac{1}{2}} - \frac{1}{24}\delta^3 f_{\frac{1}{2}} + \frac{3}{640}\delta^5 f_{\frac{1}{2}} - \ldots \right\}$ in all cases where central

differences are available we have $h = 0.4$ and:

x	δ	$-\frac{1}{24}\delta^3$	$+\frac{3}{640}\delta^5$	Total $\times \frac{1}{h}$	$f'(x)$
3·0	141 60	− 184	+ 09	0·141 42 × 2·5	= 0·353 55
3·4	129 21	− 116	+ 03	0·129 10 × 2·5	= 0·322 75
3·8	119 60	− 78	+ 02	0·119 52 × 2·5	= 0·298 80
4·2	111 86	− 55	+ 01	0·111 81 × 2·5	= 0·279 52
4·6	105 45	− 42	—	0·105 41 × 2·5	= 0·263 52
5·0	100 03	—	—		

To find $f'(5.0)$, using (11) from § 5.4.2, because the relevant central differences are not all available, we have:

$$f'_{-\frac{1}{2}} = \frac{1}{h}\left(\nabla f_0 - \frac{1}{24}\nabla^3 f_0 - \frac{1}{24}\nabla^4 f_0 \ldots \right)$$

$$\therefore f'(5.0) = \frac{1}{0.4}\left(0.100\ 03 - \frac{0.000\ 99}{24} - \frac{-\ 0.000\ 34}{24} + \text{negligible terms} \right)$$

$$= 2.5(0.100\ 03 - 0.000\ 041 + 0.000\ 014)$$

$$= 2.5 \times 0.100\ 003 = 0.250\ 01$$

Hence we obtain the following table for $f'(x)$ with its differences which are used below;

x	$f'(x)$	δ	δ^2	δ^3	δ^4	δ^5
3·0	0·353 55					
		− 3080				
3·4	0·322 75		685			
		− 2395		− 218		
3·8	0·298 80		467		79	
		− 1928		− 139		− 19
4·2	0·279 52		328		60	
		− 1600		− 79		
4·6	0·263 52		249			
		− 1351				
5·0	0·250 01					

where the fifth decimal place in values of $f'(x)$ may not be correct in all cases but greater errors would probably be introduced by rounding off to four places. (The probable error in $hf'_{\frac{1}{2}} < \frac{1}{2}$ in the fifth place and hence the probable error in $f'_{\frac{1}{2}} < (1/0.4) \times \frac{1}{2}$, i.e. $< 2\frac{1}{2}$ in the fifth place.)

For $f'(3\cdot9)$

The value of $f'(3\cdot9)$ may be evaluated by:

either (a) using formula (1) from § 5.3.1, with $x_0 = 3\cdot6$ and $p = \tfrac{3}{4}$

$$f'_p = \frac{1}{h}\left\{ \delta f_{\frac{1}{2}} + \frac{2p-1}{4}(\delta^2 f_0 + \delta^2 f_1) + \frac{3p^2 - 3p + \frac{1}{2}}{6}\,\delta^3 f_{\frac{1}{2}} + \dots \right\}$$

$$\therefore f_{\frac{3}{4}} = \frac{1}{h}\left\{ \delta f_{\frac{1}{2}} + \frac{1}{8}(\delta^2 f_0 + \delta^2 f_1) - \frac{1}{96}\,\delta^3 f_{\frac{1}{2}} - \frac{19}{768}(\delta^4 f_0 + \delta^4 f_1) \dots \right\}$$

$$\therefore f'(3\cdot9) = \frac{1}{0\cdot4}\left\{ 0\cdot119\,60 + \frac{-0\cdot017\,35}{8} - \frac{0\cdot001\,87}{96} - \frac{-0\cdot001\,45 \times 19}{768} \right\}$$

$$= 2\cdot5\,\{\,0\cdot119\,60 - 0\cdot002\,169 - 0\cdot000\,019 + 0\cdot000\,036\,\}$$

$$= 2\cdot5 \times 0\cdot117\,45$$

$$= 0\cdot2936\ \text{(to 4D)}$$

or (b) by interpolating in the table of values of $f'(x)$ using, e.g. the Bessel interpolation formula in the form:

$$f'_p = f'_0 + p\delta f'_{\frac{1}{2}} + B_2(\delta^2_m f_0 + \delta^2_m f_1) + B_3 \delta^3 f_{\frac{1}{2}}$$

as third differences are greater than 60 but fourth differences less than 1,000 Here $x_0 = 3\cdot8$, $p = 0\cdot25$, $\delta^2_m f'_0 = 0\cdot004\,52$, $\delta^2_m f'_1 = 0\cdot003\,17$ and obtaining B_2 and B_3 from Table 7 I.A.T. p. 39 we have:

$$'_p = 0\cdot298\,80 + (0\cdot25)\,(-0\cdot019\,28) + (-0\cdot046\,875)\,(0\cdot007\,69)$$

$$+ (0\cdot007\,81)\,(-0\cdot001\,39)$$

$$= 0\cdot298\,80 - 0\cdot004\,82 - 0\cdot000\,361 - 0\cdot000\,011$$

$$= 0\cdot2936\ \text{(to 4D)}$$

For $f''(4)$

The value of $f''(4)$ may be evaluated:

either (a) using formula (13), from § 5.6, with $x_0 = 4$

$$f''_0 = \frac{1}{h^2}\left\{ \delta^2 f_0 - \frac{1}{12}\,\delta^4 f_0 + \frac{1}{90}\,\delta^6 f_0 \dots \right\}$$

$$\therefore f''(4) = \frac{1}{(0\cdot4)^2}\left\{ -0\cdot007\,74 - \frac{-0\cdot000\,54}{12} + \frac{-0\cdot000\,17}{90} \dots \right\}$$

$$= 6\cdot25\,\{\,-0\cdot007\,74 + 0\cdot000\,045 - 0\cdot000\,002 \dots \}$$

$$= 6\cdot25 \times -0\cdot007\,697$$

$$= -0\cdot0481\ \text{(to 4D)}$$

or (b) using formula (2), from § 5.3.1, with $x_0 = 3.8$ in the table of $f'(x)$

$$f_{\frac{1}{2}}' = \frac{1}{h}\left\{ \delta f'_{\frac{1}{2}} - \frac{1}{24}\delta^3 f_{\frac{1}{2}} + \frac{3}{640}\delta^5 f_{\frac{1}{2}} \cdots \right\}$$

$$\therefore f''(4) = \frac{1}{0.4}\left\{ -0.019\ 28 - \frac{-0.001\ 39}{24} + \frac{-0.000\ 20 \times 3}{640} \cdots \right\}$$

$$= 2.5\ \{ -0.019\ 28 + 0.000\ 058 - 0.000\ 001 \ldots \}$$

$$= 2.5 \times -0.019\ 223$$

$$= -0.0481 \text{ (to 4D)}$$

5.7 EXAMPLES

1. If $y = e^x (1 - x)$, find the values of dy/dx and d^2y/dx^2 at $x = 0.5$ by tabulating the function at intervals of 0.1 to 4D. Check the reliability of your answers by verifying that $\dfrac{d^2y}{dx^2} + y = 2\dfrac{dy}{dx}$.

2. Given the following table of values:

x	$f(x)$	x	$f(x)$
0	1·0	0·5	0·520 574
0·1	0·900 167	0·6	0·435 358
0·2	0·801 330	0·7	0·355 782
0·3	0·704 480	0·8	0·282 644
0·4	0·601 582		

(a) Evaluate $f'(x)$ for $x = 0.15(0.1)0.65$.

(b) Evaluate $f'(0.3)$ and $f'(0.5)$ by interpolating in values of $f'(x)$ and check by finding their values from $f(x)$ directly.

(c) Find $f''(0.45)$ and $f''(0.4)$ and check accuracy by verifying that in each case $f''(x) + f(x) = 1$.

3. Given the following table of values:

x	0	0·2	0·4	0·6	0·8	1·0	1·2
$f(x)$	1·000	1·586	2·098	2·498	2·786	3·000	3·216

(a) Show that $f(x)$ is a quartic in x with given exact values.

(b) Evaluate $f'(x)$ for $x = 0(0.2)1.2$.

(c) Check your results by showing that the values obtained for $f'(x)$ satisfy a cubic in x.

(d) Evaluate $f''(x)$ for $x = 0.2(0.2)1.0$ and check your results.

4. Given the following table of exact values:

x	0	0·1	0·2	0·3	0·4	0·5
$f(x)$	7·000	7·164	7·272	7·348	7·416	7·500

(a) Calculate the values of $f'(x)$ for $x = 0.1(0.1)0.4$.

(b) Calculate the values of $f'(0.22)$ and $f''(0.25)$.

5. The table gives values of a function $y = f(x)$. Evaluate dy/dx for $x = 0.217$ to four significant figures.

x	y	x	y
0·0	0·620 545	0·25	0·632 250
0·05	0·623 265	0·30	0·634 028
0·10	0·625 793	0·35	0·635 621
0·15	0·628 133	0·40	0·637 028
0·20	0·630 285		

(A.E.B. 1966)

6. Use the table given to find $df(x)/dx$ over a suitable range. Hence determine the value of x for which $f(x)$ takes a maximum value, and give this value.

x	$f(x)$	$\delta^2 f$	x	$f(x)$	$\delta^2 f$
3·0	168·305	− 1688	4·0	199·509	− 1992
3·2	178·253	− 1782	4·2	199·925	− 1998
3·4	186·419	− 1862	4·4	198·343	− 1980
3·6	192·723	− 1926	4·6	194·781	− 1936
3·8	197·101	− 1970			

(A.E.B. 1963)

7. Given:

x	$f(x)$	x	$f(x)$
0·05	0·049 958	0·30	0·291 457
0·10	0·099 669	0·35	0·336 675
0·15	0·148 890	0·40	0·380 506
0·20	0·197 396	0·45	0·422 854
0·25	0·244 979	0·50	0·463 648

Tabulate $y = (1 + x^2)f(x)$ for $x = 0(0 \cdot 05)0 \cdot 5$ to 6D.

Evaluate $\dfrac{dy}{dx}$ for $x = 0 \cdot 1(0 \cdot 05)0 \cdot 3$.

Verify that $x \dfrac{d^2y}{dx^2} - \dfrac{dy}{dx} = \dfrac{x^2 - 1}{x^2 + 1}$ at $x = 0 \cdot 2$. (A.E.B. 1962)

8. The table gives values of a function $y = f(x)$ whose seventh differences may be neglected. Evaluate dy/dx at $x = 3 \cdot 482$ to 3D.

x	y	x	y	x	y
2·2	10·0350	3·0	21·0955	4·0	55·6082
2·4	12·0332	3·2	25·5425	4·2	67·6963
2·6	14·4737	3·4	30·9741	4·4	82·4609
2·8	17·4546	3·6	37·6082	4·6	100·4943
		3·8	45·7112		

(A.E.B. 1961)

9. Show that the values in the table satisfy $y = f(x)$ where $f(x)$ is a quintic in x

x	y	x	y
1·9	399·399	2·4	−1368·576
2·0	0	2·5	−1546·875
2·1	−395·199	2·6	−1607·424
2·2	−768·768	2·7	−1518·993
2·3	−1100·757	2·8	−1247·232

Evaluate $\dfrac{d^2y}{dx^2}$ at $x = 2 \cdot 0(0 \cdot 1)2 \cdot 7$. (A.E.B. 1960)

10.

x	$f(x)$	$\delta_{\frac{1}{2}}$	$\delta^3_{\frac{1}{2}}$	$\delta^5_{\frac{1}{2}}$	$\delta^7_{\frac{1}{2}}$
		190	−100	10	−10
0·5	1763·57				
		2413	−102	108	−12
0·6	1787·70				
		4534	4	194	−11

Using the above table find $\dfrac{\mathrm{d}f}{\mathrm{d}x}$ at $x = 0\cdot56$

and $\dfrac{\mathrm{d}^2f}{\mathrm{d}x^2}$ at $x = 0\cdot6$.

In both cases state the reliability of your answers assuming that the tabular function values are rounded off.

(Cambridge 1966)

11. Part of a table of $f(x)$ and its even order differences is given

x	$f(x)$	δ^2	δ^4
3·75	12·000 00	+ 10 21	−54 17
4·00	11·147 56	− 929 06	−53 93
4·25	9·366 06	−1922 26	−53 70

sixth and higher differences being negligible. Find as accurately as possible $(\mathrm{d}^2f/\mathrm{d}x^2)$ at $x = 4\cdot00$ assuming that the given values of $f(x)$ may be in error by $\pm\, 0\cdot000\,005$. (Cambridge 1965)

12. Correct any errors in the table of the quartic $y = f(x)$ and extend it to $x = 2\cdot0$.

x	0·8	0·9	1·0	1·1	1·2	1·3
$f(x)$	2·9966	2·5631	2·1470	1·7711	1·4606	1·2443

x	1·4	1·5	1·6	1·7
$f(x)$	1·1486	1·2095	1·4606	1·9391

Evaluate $\dfrac{\mathrm{d}y}{\mathrm{d}x}$ for $x = 1\cdot3(0\cdot1)1\cdot8$ (A.E.B. 1959)

13. Values of $\log_e(x!)$ are tabulated for integral values of x on page 307, *Chamber's Shorter Six-figure Tables*, If these values are taken to define a continuous function of x, calculate the gradient of this function at $x = 9$.

14. Tabulate values of $\log_e x$ near $x = 5$ using (a) $h = 0\cdot5$, (b) $h = 0\cdot2$, (c) $h = 0\cdot1$, (d) $h = 0\cdot02$, and compare the accuracy attainable, with each interval, in applying central difference formula to calculate $(\mathrm{d}/\mathrm{d}x)\,[\log_e x]$ at $x = 5$.

6

An Introduction to the Numerical Solution of Differential Equations

6.1 DIFFERENTIAL EQUATIONS

A differential equation is a functional relationship between two or more variables in which at least one derivative is explicit. For example

$$x\frac{\mathrm{d}y}{\mathrm{d}x} + 2y^2 = 3xy \quad \text{and} \quad 3\frac{\mathrm{d}^2s}{\mathrm{d}t^2} + 4\frac{\mathrm{d}s}{\mathrm{d}t} = \cos t$$

are differential equations.

Differential equations are classified according to the highest derivative present into first-, second-, or higher-order differential equations. The two above equations are first- and second-order respectively.

Differential equations occur widely in problems in science and engineering. For example, the equation of motion of a body falling under gravity in a resisting medium, where the resistance is proportional to the square of the velocity is $(m\mathrm{d}v/\mathrm{d}t) = m\mathrm{g} - kv^2$, (using Newton's second law of motion), a first-order differential equation connecting velocity and time. This equation may also be written as $mv(\mathrm{d}v/\mathrm{d}s) = m\mathrm{g} - kv^2$, a first-order differential equation connecting velocity and distance, or as $(m\mathrm{d}^2s/\mathrm{d}t^2) = m\mathrm{g} - k(\mathrm{d}s/\mathrm{d}t)^2$ a second-order differential equation connecting distance and time.

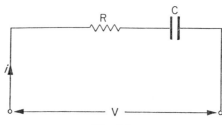

Fig. 6.1

In the circuit shown in Fig. 6.1, if v is the voltage across the capacitor C, then the current i is given by $i = C(dv/dt)$. If V is some constant voltage, switched across the complete circuit of the capacitor and the resistor R in series, then

$$V = v + iR \text{ using Ohm's Law}$$

and thus
$$V = v + RC \frac{dv}{dt}$$

which is a first-order differential equation connecting v and t.

An important group of differential equations are those of the form

$$P_n \frac{d^n y}{dx^n} + P_{n-1} \frac{d^{n-1}y}{dx^{n-1}} + \ldots\ldots + P_1 \frac{dy}{dx} + P_0 y = f(x)$$

where all the P's are functions of x only. These are called linear differential equations.

In the rest of this chapter dy/dx will usually be denoted by y' and (d^2y/dx^2) by y'', etc. The differential equation

$$\frac{d^2 y}{dx^2} - 3 \frac{dy}{dx} + y = 3x^2$$

is thus written as $y'' - 3y' + y = 3x^2$.

6.1.1 The solution of a differential equation would ideally be an equation of the form $y = f(x)$. In the simplest cases this is easily found by direct integration. For example if $y' = 6x$ then $y = 3x^2 + c$. Similarly, the equation for motion in a straight line with constant acceleration f and no resistances, $(d^2s/dt^2) = f$ or $s'' = f$, can be integrated to give $s' = ft + c_1$ and integrated again to give $s = \frac{1}{2}ft^2 + c_1 t + c_2$.

Note how arbitrary constants arise in the course of the integration so that these differential equations have an infinite number of solutions. In order to find the desired solution of a differential equation it is necessary to have some prior information about the solution. In the first example above it is sufficient to know one point of the solution (x_1, y_1) in order to determine the value of c. In the second example there are two arbitrary constants so that two facts about the solution are necessary. These could be two points (t_1, s_1) and (t_2, s_2), but in this case would probably be the initial displacement and the initial velocity such as $s = 0$ and $s' = u$ when $t = 0$ thus giving $c_2 = 0$ and $c_1 = u$ and hence $s = ut + \frac{1}{2}ft^2$. In general it is necessary to know the same number of facts about the solution as the order of the differential equation. In the case of second-order differential equations these may be either the co-ordinates of two points or the values of the function and its first derivative at one point. The first is called a two point boundary problem and the second a one point boundary problem. The numerical solution of these two types of problem is quite different and we limit discussion to the cases in which the given conditions occur at only one point, i.e. one point boundary problems, (also known as initial value problems.)

E

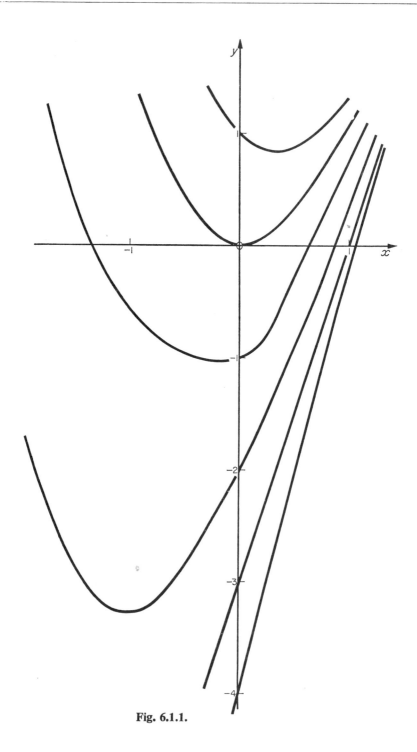

Fig. 6.1.1.

Some solutions of the first order equation $y' + y = 3x$ with different initial conditions are illustrated in Fig. 6.1.1.

Note that the solutions of a differential equation form a family of curves. The particular given condition means that one of this family becomes the solution of the particular problem.

Other equations in appropriate circumstances also result in families of solutions.

Although most differential equations cannot be solved by direct integration various analytical techniques exist which are applicable to several types of equation. For example such an analytical solution of the equation $y' + y = 3x$ is discussed in § 6.2.10. Details of other methods will be found in most standard textbooks on Pure Mathematics. However, even though an analytical solution exists, it may be very difficult to obtain or may result in an integral which must be tabulated by a numerical process. In such cases a numerical solution of the differential equation is preferable, that is, a method which results in a table of values of the solution function for some desired range of values of the independent variable.

In this Chapter we are only able to give a brief account of some of the simpler numerical methods. In fact some of the methods most extensively used with digital computers (such as the Runge–Kutta method) are not discussed at all because they rely on mathematics which the reader has probably not yet studied. It is hoped, however that the following methods will be of interest and will stimulate a desire for a deeper study of the numerical methods for the solution of differential equations.

6.2 THE SOLUTION OF FIRST ORDER DIFFERENTIAL EQUATIONS

Three methods for the solution of first order differential equations are presented: in § 6.2.1 a graphical method, in § 6.2.3 Fox and Goodwin's method for linear equations and in § 6.2.7 a step-by-step method using central differences.

6.2.1 A graphical method of solution of first-order differential equations using a step-by-step method has 'build-up' error.

This method is illustrated by its application to the solution of the equation $y' + y = 3x$ in the range $x = 0$ to $x = 1$ with $y = 1$ when $x = 0$.

Any first-order differential equation can be written in the form $y' = f(x, y)$ giving an expression for the gradient in terms of x and y. Thus in this example $y' = 3x - y$.

The given initial point (0, 1) is plotted (P_0 in Fig. 6.2.1) and the gradient of the solution curve at this point evaluated by substitution of these values of x and y in the above expression giving here $y' = -1$. A short straight line with this gradient is now drawn through the point P_0 to cross the line $x = 0 \cdot 1$ giving there the point P_1 $(0 \cdot 1, 0 \cdot 9)$. This point lies very near to the solution curve.

The gradient y^1 at P_1 is now calculated and a line with this gradient drawn through P_1 to cross the line $x = 0.2$ giving P_2. This process is now repeated at intervals of 0.1 in x up to $x = 1$. The resulting graph is shown in figure 6.2.1, in which the gradients were rounded off to one decimal place.

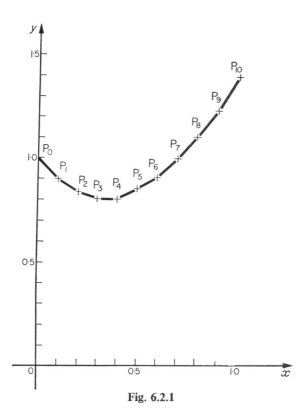

Fig. 6.2.1

Accuracy factors

Several factors affect the accuracy of this solution. The working was carried out at intervals of 0.1 in x, but if a smaller interval is chosen then successive points will lie nearer together and nearer to the true solution. The size of the interval used depends to some extent on the size of the scale used in the drawing. This should be as large as possible depending on the range of values sought and the graph paper available. The size of scale also affects the accuracy with which the gradients can be drawn.

This solution advances one step at a time, obtaining each point from the previous one. Such a method is described as a step-by-step method. In all step-by-step methods the errors in the values obtained increase as the solution is developed further away from the starting point. For example in this solution at $x = 0.1$, $y = 0.90$ compared with the correct value of 0.92 to 2D, whereas at $x = 1·0$, $y = 1·42$ compared with the correct value of $1·47$ to 2D. That is, the error in P_1 is about $0·02$ and in P_{10} about $0·05$. In many cases this 'build-up' error will be greater than this. Used with caution though, over a short range, this method can certainly give qualitative information on the form of the solution and indeed give quantitative information of limited accuracy. It is capable of further refinement to increase the accuracy of the solution. (e.g. See Redish p. 112).

6.2.2 Examples

Sketch the solutions of:

1. $y' - 2xy = 3x^2 - 1$ subject to $y = -1$ at $x = 0$ over the range $x = -0·6(0·1)0·5$.

2. $y' + y \sin x = 4x^3$ subject to $y = 0·5$ at $x = 0$ over the range $x = -0·8(0·1)0·8$.

3. $y' = 2xy^2 + x^2$ subject to $y = 0·1$ at $x = 0$ over the range $x = -1·0(0·1)1·0$.

4. $y' = 4x^2y^2$ subject to $y = 0·25$ at $x = 0$ over the range $x = -0·8(0·1)0·8$.

5. $y' + y \sin x = \cos x$ subject to $y = 0$ at $x = 0$ over the range $x = 0(0·1)1·2$.

6.2.3 Fox and Goodwin's method of deferred correction is an iterative method of solution of differential equations, in which approximate values of the required function are first found over the whole required range before subsequent improvement in accuracy.

The method is most suitable for application to linear differential equations, which in the first-order case may be written in the form $y' + Py = Q$ where P and Q are functions of x. Only this type is discussed below.

The basis of this method is a differentiation formula obtained from equation (4) Chapter 5, §5.3.2. This is:

$$hf'_0 = \tfrac{1}{2}(\delta f_{-\frac{1}{2}} + \delta f_{\frac{1}{2}}) - \tfrac{1}{12}(\delta^3 f_{-\frac{1}{2}} + \delta^3 f_{\frac{1}{2}}) + \tfrac{1}{60}(\delta^5 f_{-\frac{1}{2}} + \delta^5 f_{\frac{1}{2}}) - \cdots$$

thus $hy'_0 = \tfrac{1}{2}(\delta y_{-\frac{1}{2}} + \delta y_{\frac{1}{2}}) - \tfrac{1}{12}(\delta^3 y_{-\frac{1}{2}} + \delta^3 y_{\frac{1}{2}}) + \tfrac{1}{60}(\delta^5 y_{-\frac{1}{2}} + \delta^5 y_{\frac{1}{2}}) - \cdots$

and $hy'_1 = \tfrac{1}{2}(\delta y_{\frac{1}{2}} + \delta y_{1\frac{1}{2}}) - \tfrac{1}{12}(\delta^3 y_{\frac{1}{2}} + \delta^3 y_{1\frac{1}{2}}) + \tfrac{1}{60}(\delta^5 y_{\frac{1}{2}} + \delta^5 y_{1\frac{1}{2}}) - \cdots$

giving $h(y'_0 + y'_1) = \tfrac{1}{2}(\delta y_{-\frac{1}{2}} + 2\delta y_{\frac{1}{2}} + \delta y_{1\frac{1}{2}}) - \tfrac{1}{12}(\delta^3 y_{-\frac{1}{2}} + 2\delta^3 y_{\frac{1}{2}} + \delta^3 y_{1\frac{1}{2}})$

$$+ \tfrac{1}{60}(\delta^5 y_{-\frac{1}{2}} + 2\delta^5 y_{\frac{1}{2}} + \delta^5 y_{1\frac{1}{2}}) - \cdots.$$

but as $\delta^2 y_0 = \delta y_{\frac{1}{2}} - \delta y_{-\frac{1}{2}}$ then $\delta y_{-\frac{1}{2}} = \delta y_{\frac{1}{2}} - \delta^2 y_0$ and similarly $\delta y_{1\frac{1}{2}} = \delta y_{\frac{1}{2}} + \delta^2 y_1$

so that $\delta y_{-\frac{1}{2}} + 2\delta y_{\frac{1}{2}} + \delta y_{1\frac{1}{2}} = 4\delta y_{\frac{1}{2}} + \delta^2 y_1 - \delta^2 y_0 = 4\delta y_{\frac{1}{2}} + \delta^3 y_{\frac{1}{2}}$

and similarly for the other terms giving

$h(y'_0 + y'_1) = \tfrac{1}{2}(4\delta y_{\frac{1}{2}} + \delta^3 y_{\frac{1}{2}}) - \tfrac{1}{12}(4\delta^3 y_{\frac{1}{2}} + \delta^5 y_{\frac{1}{2}}) + \tfrac{1}{60}(4\delta^5 y_{\frac{1}{2}} + \delta^7 y_{\frac{1}{2}}) - \cdots$

which on simplifying gives

$$h(y_0' + y_1') = 2\delta y_{\frac{1}{2}} + \tfrac{1}{6}\delta^3 y_{\frac{1}{2}} - \tfrac{1}{60}\delta^5 y_{\frac{1}{2}} + \ldots$$

or $\tfrac{1}{2}h(y_0' + y_1') = y_1 - y_0 + \tfrac{1}{12}\delta^3 y_{\frac{1}{2}} - \tfrac{1}{120}\delta^5 y_{\frac{1}{2}} + \ldots$

Assuming, that one value of y, y_0 at x_0 is known and, that the value y_1 at $x_1 = x_0 + h$ is sought, we obtain on substituting these values in the original equation

$$y_0' + P_0 y_0 = Q_0$$

and

$$y_1' + P_1 y_1 = Q_1$$

Adding these two equations gives

$$y_0' + y_1' + P_0 y_0 + P_1 y_1 = Q_0 + Q_1$$

and multiplying by $\tfrac{1}{2}h$

$$\tfrac{1}{2}h(y_0' + y_1') + \tfrac{1}{2}h\, P_0 y_0 + \tfrac{1}{2}h\, P_1 y_1 = \tfrac{1}{2}h(Q_0 + Q_1)$$

and we have:

$$y_1 - y_0 + \tfrac{1}{12}\delta^3 y_{\frac{1}{2}} - \tfrac{1}{120}\delta^5 y_{\frac{1}{2}} + \ldots + \tfrac{1}{2}h\, P_0 y_0 + \tfrac{1}{2}h\, P_1 y_1 = \tfrac{1}{2}h(Q_0 + Q_1)$$

or $y_1(1 + \tfrac{1}{2}h\, P_1) - y_0(1 - \tfrac{1}{2}h\, P_0) + C(y_{\frac{1}{2}}) = \tfrac{1}{2}h(Q_0 + Q_1)$

where $\quad C(y_{\frac{1}{2}}) = \tfrac{1}{12}\delta^3 y_{\frac{1}{2}} - \tfrac{1}{120}\delta^5 y_{\frac{1}{2}} + \ldots$

giving the recurrence relation:

$$y_1 = \frac{y_0(1 - \tfrac{1}{2}h\, P_0) + \tfrac{1}{2}h(Q_0 + Q_1) - C(y_{\frac{1}{2}})}{1 + \tfrac{1}{2}h\, P_1}$$

The term $C(y_{\frac{1}{2}})$ is regarded as a 'correction factor' and is ignored for the first approximation so that then everything on the right-hand side is known and y_1 can be found.

For a second approximation to y_1 it is necessary to obtain an approximate value of $C(y_{\frac{1}{2}})$. This requires approximate values of $\delta^3 y_{\frac{1}{2}}$, $\delta^5 y_{\frac{1}{2}}$, etc., but in simple cases only that of $\delta^3 y_{\frac{1}{2}}$ may be necessary as the contribution of the higher differences may be insignificant.

It is thus necessary to obtain a table of values of first approximations to y from which values of the differences may be found. This table is obtained by repeated application of the above recurrence relation, neglecting $C(y_{\frac{1}{2}})$. i.e. The values of y_0, P_0, Q_0, P_1 and Q_1 are substituted to obtain y_1 and then the values of y_1, P_1, Q_1, P_2 and Q_2 to obtain y_2. The process is repeated as far as required to give y_3, y_4, etc.

So that all the required differences may be found it is sometimes necessary to extend the table backwards. The above recurrence relation may be rearranged to give y_0 in terms of y_1 as

$$y_0(1 - \tfrac{1}{2}h\, P_0) = y_1(1 + \tfrac{1}{2}h\, P_1) - \tfrac{1}{2}h(Q_0 + Q_1) + C(y_{\frac{1}{2}})$$

which for this purpose is more conveniently written as

$$y_{-1} = \frac{y_0(1 + \frac{1}{2}h\,P_0) - \frac{1}{2}h(Q_{-1} + Q_0) + C(y_{-\frac{1}{2}})}{1 - \frac{1}{2}h\,P_{-1}}$$

Worked example

Consider again the solution of $y' + y = 3x$, where $y = 1$ when $x = 0$, in the range $x = 0$ to $x = 0\cdot4$ to 4D. Here $P = 1$ and $Q = 3x$.

First approximation, y_1

Using an interval of $0\cdot1$ the recurrence relation for the first approximation is:

$$y_1 = \frac{y_0(1 - 0\cdot05P_0) + 0\cdot05(Q_0 + Q_1)}{1 + 0\cdot05P_1}$$

which is simpler to evaluate if the numerator and the denominator of the right-hand side are multiplied by 20 thus giving:

$$y_1 = \frac{y_0(20 - P_0) + Q_0 + Q_1}{20 + P_1}$$

It is also useful to tabulate P and Q over the required range and to note these values alongside the working as shown so that in most cases the value of y_1 may be obtained by a continuous process on the calculating machine.

For the first step of the solution we have $x_0 = 0$, $y_0 = 1$ and $x_1 = 0\cdot1$ giving $P_0 = 1$, $P_1 = 1$, $Q_0 = 0$ and $Q_1 = 0\cdot3$ so that on substitution in the recurrence relation we obtain $y_1 = 0\cdot919\,048$ to 6D, i.e. carrying two guarding figures, (although in many cases it is sufficient to work with fewer figures in the first approximation.)

To find the value of y at $x = 0\cdot2$ we have $x_0 = 0\cdot1$, $y_0 = 0\cdot919\,048$ and $x_1 = 0\cdot2$ giving $P_0 = 1$, $P_1 = 1$, $Q_0 = 0\cdot3$ and $Q_1 = 0\cdot6$. The operations involved in this evaluation are shown below:

S.R. 5D C.R. 6D Acc. 11D	S.R.	C.R.	Acc.
Set $20 - P_0 = 19$	19·000 00	0	0
Mult. by $y_0 = 0\cdot919\,048$			
return carriage to place 7	19·000 00	0·919 048	17·461 912 000 00
Clear C.R. set and add			
$\quad Q_0 = 0\cdot3$	0·300 00	1·000 000	17·761 912 000 00
Clear C.R. set and add			
$\quad Q_1 = 0\cdot6$	0·600 00	1·000 000	18·361 912 000 00
Clear C.R. set $20 + P_0 = 21$	21·000 00	0	18·361 912 000 00
Complete division	21·000 00	0·874 376	0·000 016 000 00

Inspection of the remainder (16) shows that the result should be correctly rounded off to 0·874 377.

Note how the careful choice of decimal places in the registers ensured that no change had to be made in these for the division.

Further values of y are found in the same way and are tabulated and differenced as shown below. y_I refers to the first approximation to y.

Table of values of y_I

x	y_I					P	Q
0	1					1	0
		$-80\,952$					
0·1	0·919 048		36 281			1	0·3
		$-44\,671$		$-3\,455$			
0·2	0·874 377		32 826			1	0·6
		$-11\,845$		$-3\,127$			
0·3	0·862 532		29 699			1	0·9
		$+17\,854$					
0·4	0·880 386					1	1·2

It will now be seen that there are insufficient third differences to proceed to the second approximation, so that it is necessary to calculate y_I at $x = 0·5$ as before, and at $x = -0·1$ using the backwards form of the recurrence relation which, neglecting $C(y_{-\frac{1}{2}})$ and simplifying, is here:

$$y_{-1} = \frac{y_0(20 + P_0) - Q_{-1} - Q_0}{20 - P_{-1}}$$

The extended table of values of y_I is then as follows:

x	y_I				$\frac{5}{3}\delta^3 y_{\frac{1}{2}}$	P	Q
$-0·1$	1·121 053					1	$-0·3$
		$-121\,053$					
0	1		40 101			1	0
		$-80\,952$		$-3\,820$	$-6\,367$		
0·1	0·919 048		36 281			1	0·3
		$-44\,671$		$-3\,455$	$-5\,758$		
0·2	0·874 377		32 826			1	0·6
		$-11\,845$		$-3\,127$	$-5\,212$		
0·3	0·862 532		29 699			1	0·9
		$+17\,854$		$-2\,828$	$-4\,713$		
0·4	0·880 386		26 871			1	1·2
		44 725					
0·5	0·925 111					1	1·5

Second approximation y_{II}

With some third differences now available we may obtain the second approximation by including the first term of the correction factor in the recurrence relation which is thus:

$$y_1 = \frac{y_0(1 - 0{\cdot}05P_0) + 0{\cdot}05(Q_0 + Q_1) - \frac{1}{12}\delta^3 y_{\frac{1}{2}}}{1 + 0{\cdot}05P_1}$$

or

$$y_1 = \frac{y_0(20 - P_0) + Q_0 + Q_1 - \frac{5}{3}\delta^3 y_{\frac{1}{2}}}{20 + P_1}$$

$\frac{5}{3}\delta^3 y_{\frac{1}{2}}$ is tabulated as above in order to simplify the working.

The solution again begins from $x_0 = 0$ and $y_0 = 1$ and is continued using the values of y calculated in the *second* approximation (y_{II}). The values of y obtained above under y_I are not directly used in the second approximation but only the resulting values of the third differences in the term $\frac{5}{3}\delta^3 y_{\frac{1}{2}}$.

Table including values of y_{II} *and* y_{III}

The table of values of y_{II} and its differences are now added to the above table as follows on page 128.

It will be seen that the change in value of $\frac{5}{3}\delta^3 y_{\frac{1}{2}}$ will not cause any change in the fifth decimal place in the values of y_{II}. Calculating the fifth differences from the table of values of y_I shows that their contribution is negligible. The table of values of y_{II} may be rounded off correctly to 4D. However, in order to confirm this, it may be useful to obtain y_{III}, with estimates for the missing values of $\frac{5}{3}\delta^3 y_{\frac{1}{2}}$ in order to save all the working necessary in calculating these exactly. Such estimates and the resulting table of values of y_{III} are shown on page 128. The values of y_{III} agree with those of y_{II} to the required 4D so that the solution is:

x	0	0·1	0·2	0·3	0·4
y	1	0·9194	0·8749	0·8633	0·8813

The number of approximations necessary when using this method depends on the magnitude of $C(y_{\frac{1}{2}})$ which in turn depends on the magnitude of the third and fifth differences. If these are found to be large at the first approximation then it is probably quicker to start again with a smaller interval, thus speeding up the convergence.

6.2.4 Examples

1 Integrate $y' - 2xy = 3x^2 - 1$ subject to $y = -1$ at $x = 0$, over the range $x = 0(0{\cdot}1)0{\cdot}5$ to 4D.

x	y_I	$\frac{5}{9}\delta^3 y_{\frac12}$	y_{II}	$\frac{5}{9}\delta^3 y_{\frac12}$	y_{III}	P	Q
− 0·1	1·121 053					1	− 0·3
	− 121 053						
0	1		1		1	1	0
	− 80 952 40 101 − 3 820	− 6 367	− 80 649 36 224 − 3 449	(− 6360)			
0·1	0·919 048		0·919 351		0·919 350	1	0·3
	− 44 671 36 281 − 3 455	− 5 758	− 44 425 32 775 − 3 117	− 5 748			
0·2	0·874 377		0·874 926		0·874 924	1	0·6
	− 11 845 32 826 − 3 127	− 5 212	− 11 650 29 658	− 5 195			
0·3	0·862 532		0·863 276		0·863 274	1	0·9
	+ 17 854 29 699 − 2 828	− 4 713	+ 18 008	(− 4700)			
0·4	0·880 386		0·881 284		0·881 281	1	1·2
	44 725 26 871						
0·5	0·925 111					1	1·5

2. Integrate $y' + y \sin x = 4x^3$ subject to $y = 0.5$ at $x = 0$ over the range $x = 0(0.1)0.5$ to 3D.

3. Integrate $y' + 3x^2y = 2x^2 + 3x - 1$ subject to $y = 0$ at $x = 0$ over the range $x = 0(0.1)0.6$ to 3D.

4. Integrate $y' + 2xy = \cos x^2$ subject to $y = 0.5$ at $x = 0$ over the range $x = 0(0.1)0.5$ to 4D.

6.2.5 A series solution of differential equations is described next because of its usefulness in beginning the step-by-step solution which is discussed later.

A power series of the form $y = a_0 + a_1x + a_2x^2 + a_3x^3 + a_4x^4 + \ldots$ is sought which approximates to y over a small range of values of x. It is here assumed that the value of y when $x = 0$ is given, so that series solutions around $x = 0$ will be considered. Series solutions around other values of x are possible but they are rather more involved and are not discussed below.

Worked example

Find the series solution of $y' + y = 3x$ with $y = 1$ when $x = 0$.

Let $y = a_0 + a_1x + a_2x^2 + a_3x^3 + a_4x^4 + \ldots$

Since $y = 1$ when $x = 0$ we have $a_0 = 1$.

Differentiating the series term by term gives:

$y' = a_1 + 2a_2x + 3a_3x^2 + 4a_4x^3 + \ldots$ but $y' = 3x - y$, from the original equation, so that $y' = -1$ at $x = 0$ and thus $a_1 = -1$.

Differentiating the series again gives:

$$y'' = 2a_2 + 3.2a_3x + 4.3a_4x^2 + \ldots$$

but $y'' = 3 - y'$, on differentiating the original equation,

so that $y'' = 4$ at $x = 0$ and thus $a_2 = 2$.

Differentiating again gives:

$$y''' = 3.2a_3 + 4.3.2a_4x + \ldots \text{ and } y''' = -y''$$

so that $y''' = -4$ at $x = 0$ and thus $a_3 = -\frac{2}{3}$.

This process of repeatedly differentiating the power series and the original equation is continued until sufficient terms of the series are known. Here we obtain:

$$y = 1 - x + 2x^2 - \tfrac{2}{3}x^3 + \tfrac{1}{6}x^4 - \tfrac{1}{30}x^5 + \ldots$$

Values of y around $x = 0$ may now be found by substitution in this series.

If the differential equation does not contain any powers of y higher than the first then the following method of comparing coefficients may be used. In these cases this method will be found to be quicker as it only involves one differentiation.

Let $y = a_0 + a_1x + a_2x^2 + a_3x^3 + a_4x^4 + \ldots$

Since $y = 1$ when $x = 0$ we have $a_0 = 1$.

Differentiating the series term by term gives

$$y' = a_1 + 2a_2x + 3a_3x^2 + 4a_4x^3 + \ldots$$

and on substitution in the original differential equation we have:

$$(a_1 + 2a_2x + 3a_3x^2 + 4a_4x^3 + \ldots) +$$

$$(1 + a_1x + a_2x^2 + a_3x^3 + a_4x^4 + \ldots) = 3x.$$

The coefficients in the series are found by comparing the coefficients of powers of x in this equation.

Thus $a_1 + 1 = 0$ so that $a_1 = -1$

$2a_2 + a_1 = 3$ so that $a_2 = 2$

$3a_3 + a_2 = 0$ so that $a_3 = -\frac{2}{3}$ and so on

The required series solution is:

$$y = 1 - x + 2x^2 - \tfrac{2}{3}x^3 + \tfrac{1}{6}x^4 - \tfrac{1}{30}x^5 + \ldots$$

This method can still be used when functions of x other than polynomials are present by using their power series expansions. These two methods can be similarly applied to the solution of second-order differential equations.

6.2.6 Examples

Find the series solutions of the following differential equations up to the e rm in x^6.

1. $y' - x^2y = 2x + 4$ with $y = 0$ when $x = 0$.

2. $y' = 3x^2y$ with $y = -1$ when $x = 0$.

3. $y' + xy = \sin x$ with $y = 1$ when $x = 0$.

4. $y' + y^2 = x^2$ with $y = 1$ when $x = 0$.

5. $x^2y' = 3y + x^2$ with $y = 0$ when $x = 0$.

6.2.7 Step-by-step solution using central differences and a series start

This is again an iterative method but each value of the solution is calculated to the required accuracy before proceeding to the next point.

It is based on the use of the integration formula, equation (5) Chapter 4, § 4.4, which is

$$\int_{x_0}^{x_1} f(x)\,dx = h\left\{ \frac{f_0 + f_1}{2} - \frac{\delta^2 f_0 + \delta^2 f_1}{24} + \frac{11}{1440}(\delta^4 f_0 + \delta^4 f_1) \dots \right\}$$

Putting $f = y'$ this gives

$$\int_{x_0}^{x_1} y'\,dx = \frac{h}{2}\left\{ (y_0 + y_1') - \frac{1}{12}(\delta^2 y_0' + \delta^2 y_1') + \frac{11}{720}(\delta^4 y_0' + \delta^4 y_1') \dots \right\}$$

But $\int_{x_0}^{x_1} y'\,dx = [y]_{x_0}^{x_1}$

$$= y_1 - y_0$$

$$= \delta y_{\frac{1}{2}}$$

and thus

$$\delta y_{\frac{1}{2}} = \frac{h}{2}\left\{ (y_0' + y_1') - \frac{1}{12}(\delta^2 y_0' + \delta^2 y_1') + \frac{11}{720}(\delta^4 y_0' + \delta^4 y_1') + \dots \right\}$$

Given y_0, and having found $\delta y_{\frac{1}{2}}$ from this equation, y_1 is immediately obtained.

In order to truncate this equation after the term in second differences, the value of h is chosen so that the other terms may be neglected. It is thus necessary that

$$\frac{11h}{1440}(\delta^4 y_0' + \delta^4 y_1') < \tfrac{1}{2}$$

in terms of the last decimal place.

This gives approximately

$$\frac{11h}{720}\delta^4 y_0' < \tfrac{1}{2} \text{ or } h\delta^4 y_0' < 32.$$

With this condition satisfied $\delta y_{\frac{1}{2}}$ may be found from the equation

$$\delta y_{\frac{1}{2}} = \tfrac{1}{2}h\{ (y_0' + y_1') - 0\cdot08\dot{3}(\delta^2 y_0' + \delta^2 y_1') \}.$$

However, in order to use this equation it is necessary to know the values of y_0', y_1', $\delta^2 y_0'$ and $\delta^2 y_1'$. A few values of y' must therefore be first obtained. One method is to find the series solution and to use it to calculate a few values of y centred on $x = 0$. Values of y' may then be found by substitution in the differential equation. Usually five values of y' are sufficient giving one fourth difference $\delta^4 y'$ from which the chosen value of h may be checked against the condition given above. (It is preferable to choose values centred on $x = 0$ rather than values beginning from $x = 0$, because then only small values of x are involved and fewer terms of the series are significant.)

Worked example

The solution of $y' + y = 3x$ in the range $x = 0$ to $x = 0.4$ with the initial condition $y = 1$ when $x = 0$.

From § 6.2.5 this has the series solution:

$$y = 1 - x + 2x^2 - \tfrac{2}{3}x^2 + \tfrac{1}{6}x^4 - \tfrac{1}{30}x^5 + \ldots$$

This series is used to calculate the values of y for $x = -0.10(0.05)0.10$ and these are substituted in the original equation to obtain the corresponding values of y'. These values are tabulated and differenced as shown below. A check may be used on the series solution by differentiating this and calculating a value of y' and comparing this with that obtained using $y' = 3x - y$. An extra term in the series will probably be necessary to obtain y' to the same accuracy. Six decimal places are retained in the working so that the solution may be obtained correct to 4D, that is, two guarding figures are carried.

x	$y' = 3x - y$			y		
-0.10	$-1.420\,684$			$1.120\,684$		
		$215\,600$			$-65\,600$	
-0.05	$-1.205\,084$	$-10\,516$		$1.055\,084$	$10\,516$	
		$205\,084$	514		$-55\,084$	-514
0	-1		$-10\,002 - 26$	1	$10\,002$	26
		$195\,082$	488		$-45\,082$	-488
0.05	$-0.804\,918$	$-9\,514$		$0.954\,918$	$9\,514$	
		$185\,568$			$-35\,568$	
0.10	$-0.619\,350$			$0.919\,350$		

In order to apply the truncated integration formula it is necessary that approximately $h \times 26 < 32$, and thus $h < 1.2$, which shows that an interval of 0.05 is satisfactory. However, if y' is tabulated at an interval of 0.1 the corresponding fourth difference is 398, and this leads to the condition that $h < 0.08$ showing that this interval would not be suitable for this particular solution. If the fourth differences increase in magnitude as the solution proceeds this value will have to be recalculated, and it may be necessary to reduce the size of interval in the later part of the working.

Estimation—correction process

In order to obtain y_1, the value of y at $x = 0.15$, $\delta y_{\frac{1}{2}}$ is calculated from:

$$\delta y_{\frac{1}{2}} = \tfrac{1}{2}h\left\{ (y_0' + y_1') - 0.083\,(\delta^2 y_0' + \delta^2 y_1') \right\}.$$

However, only y_0' is known, but approximations can be made to the values of y_1', $\delta^2 y_0'$ and $\delta^2 y_1'$ by estimating suitable values for the next two fourth differences, and extending the difference table from these. From this calculated value of $\delta y_{\frac{1}{2}}$ the value of y_1 is found.

Using the differential equation a value of y_1' is obtained, which is a second approximation to its value, from which the estimated differences may be modified, and $\delta y_{\frac{1}{2}}$ recalculated as a check repeating the correction process if necessary. The only fourth difference of y' known from the series start is -26 and, as the differences appear to be decreasing in magnitude, -25 is chosen as an approximation to the next two fourth differences (shown in brackets below). y_1' and the necessary second differences are calculated from these and the existing differences as shown in the table below.

x	$y' = 3x - y$					y				
0·10	−1·420 684					1·120 684				
		215 600					−65 600			
0·05	−1·205 084		−10 516			1·055 084		10 516		
		205 084		514			−55 084		−514	
0	−1		−10 002		−26	1		10 002		26
		195 082		488			−45 082		−488	
0·05	−0·804 918		−9 514		(−25)	0·954 918		9 514		
		185 568		463			−35 568			
0·10	−0·619 350		−9 051		(−25)	0·919 350				
		176 517		438			−26 518			
0·15	−0·442 833		−8 613			0·892 832				

Thus here $y_0' = -0.619\,350$, $y_1' = -0.442\,833$, $\delta^2 y_0' = -0.009\,051$ and $\delta^2 y_1' = -0.008\,613$ giving $\delta y_{\frac{1}{2}} = -0.026\,518$ and thus $y_1 = 0.892\,832$ as shown above. Substitution in $y' = 3x - y$ now gives $y_1' = -0.442\,832$. The previous value of y_1' is now replaced by this value and the backward differences consequently modified. This gives a new value of the first estimated fourth difference of -24 so that the next fourth difference is now estimated as being -22. These modified values are shown below.

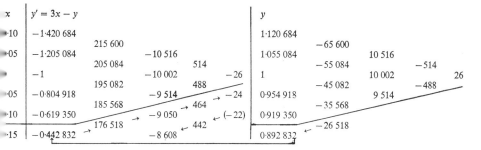

x	$y' = 3x - y$					y				
·10	−1·420 684					1·120 684				
		215 600					−65 600			
·05	−1·205 084		−10 516			1·055 084		10 516		
		205 084		514			−55 084		−514	
	−1		−10 002		−26	1		10 002		26
		195 082		488			−45 082		−488	
·05	−0·804 918		−9 514		−24	0·954 918		9 514		
		185 568		464			−35 568			
·10	−0·619 350		−9 050		(−22)	0·919 350				
		176 518		442			−26 518			
·15	−0·442 832		−8 608			0·892 832				

$\delta y_{\frac{1}{2}}$ is recalculated as a check, further modifications to the value of y_1' and the differences of y' being made if necessary. In this case $\delta y_{\frac{1}{2}}$ does not change in value so that the value of y at $x = 0.20$ may now be sought. The next fourth difference of y' may be estimated as -20 and the above process repeated.

The reader should continue as far as $x = 0.40$ and compare his final table with that shown. (It is preferable to work in pencil so that figures may be neatly changed as the solution proceeds.)

Differencing the table of values of y as well as that of y' serves as a further check.

x	$y' = 3x - y$					y			
-0.10	$-1.420\,684$					$1.120\,684$			
		$215\,600$					$-65\,600$		$10\,516$
-0.05	$-1.205\,084$		$-10\,516$			$1.055\,084$		$-55\,084$	
		$205\,084$		514				$10\,002$	-514
0	-1		$-10\,002$		-26	1		$-45\,082$	26
		$195\,082$		488				$9\,514$	-488
0.05	$-0.804\,918$		$-9\,514$		-24	$0.954\,918$		$-35\,568$	24
		$185\,568$		464				$9\,050$	-464
0.10	$-0.619\,350$		$-9\,050$		-23	$0.919\,350$		$-26\,518$	22
		$176\,518$		441				$8\,609$	-441
0.15	$-0.442\,832$		$-8\,609$		-21	$0.892\,832$		$-17\,909$	2
		$167\,909$		420				$8\,189$	-420
0.20	$-0.274\,923$		$-8\,189$		-21	$0.874\,923$		$-9\,720$	2
		$159\,720$		399				$7\,790$	-399
0.25	$-0.115\,203$		$-7\,790$		-18	$0.865\,203$		$-1\,930$	1
		$151\,930$		381				$7\,409$	-381
0.30	$+0.036\,727$		$-7\,409$		-21	$0.863\,273$		$+5\,479$	2
		$144\,521$		360				$7\,049$	-360
0.35	$0.181\,248$		$-7\,049$		(-21)	$0.868\,752$		$12\,528$	
		$137\,472$		(339)					
0.40	$0.318\,720$		$(-6\,710)$			$0.881\,280$			

Thus correct to 4D the required solution is:

x	0	0.05	0.10	0.15	0.20	0.25	0.30	0.35	0.40
y	1	0.9549	0.9194	0.8928	0.8749	0.8652	0.8633	0.8688	0.8813

The above method is but one of a number of such step-by-step methods. In some of these, approximations to certain of the values, which have here been estimated, are obtained by the use of appropriate integration formulae (often referred to as 'predictor' formulae.) The interested reader should consult one of the more advanced books given in the bibliography.

6.2.8 Examples

1. Integrate $y' + xy = 3x^2$ subject to $y = 0$ at $x = 0$ over the range $x = 0(0.05)0.30$ to 4D.

2. Integrate $y' + 3x^2y = 2x^2 + 3x - 1$ subject to $y = 0$ at $x = 0$ over the range $x = 0(0.05)0.30$ to 3D.

3. Using a suitable interval integrate $y' - 2xy = 3x^2 - 1$, subject to $y = -1$ at $x = 0$, and obtain a table of values of y over the range $x = 0(0.1)0.5$ to 4D.

4. Integrate $y' = 2xy^2 + x^2$ subject to $y = 0.1$ at $x = 0$ over the range $x = 0(0.1)1.0$ to 3D.

5. Integrate $y' = 4x^2y^2$ subject to $y = 0.25$ at $x = 0$ over the range $x = 0(0.05)0.40$ to 4D.

6.2.9   Comparison of step-by-step and Fox and Goodwin's methods

Fox and Goodwin's method has two main advantages over the step-by-step method. Firstly a larger interval may be used, thus speeding up the calculation. Secondly it uses only the given initial values of x and y and does not require any special starting technique. It is preferable to use Fox and Goodwin's method for the solution of first-order linear differential equations, and the step-by-step method for other types of first-order differential equations. Other methods have certain advantages over these two in appropriate circumstances. For example, different methods might be used with a digital computer.

6.2.10   Analytical solution of $y' + y = 3x$

Linear first-order differential equations, that is, those of the form $y' + Py = Q$, can be integrated directly on multiplication by an 'integrating factor' of $e^{\int P dx}$. As the coefficient of y is here 1 the equation is integrated by first multiplying by e^x as follows:

$$e^x \frac{dy}{dx} + e^x y = 3x e^x$$

$$\text{so that } \frac{d}{dx}(e^x y) = 3x e^x$$

$$\text{and hence } e^x y = \int 3x e^x dx + c$$

$$= 3x e^x - 3e^x + c$$

$$\text{so that } \quad y = 3x - 3 + c e^{-x}$$

But here $y = 1$ when $x = 0$ giving $c = 4$ so that

$$y = 3x - 3 + 4e^{-x}$$

Any values of y which are required are obtained by substitution in this equation. For example on substituting $x = 0.3$ we obtain $y = 0.8633$ correct to 4D, confirming the value previously obtained by numerical methods.

In this example the analytical solution is much quicker than the above numerical solutions. In general a numerical solution should only be used after analytical solutions have been investigated and found to be more difficult. For example, consider the analytical solution of $(dy/dx) + e^x y = e^{-x}$.
The integrating factor is e^{e^x}, and multiplying the equation by this gives

$$e^{e^x} \frac{dy}{dx} + e^{e^x} e^x y = e^{e^x} e^{-x}$$

$$\text{that is } \quad \frac{d}{dx}(e^{e^x} y) = e^{e^x} e^{-x}$$

$$e^{e^x} y = \int e^{e^x - x} dx + c$$

which is rather difficult to evaluate and so a numerical method of solution should be used.

F

The analytical solution is further complicated if there are higher powers of y in the original equation.

6.3 THE SOLUTION OF SECOND-ORDER DIFFERENTIAL EQUATIONS

The same principles as those for first-order differential equations are used in the solution of second-order differential equations, and there are similar step-by-step and deferred correction methods. Only a step-by-step method for the solution of the equation $y'' = f(x, y)$, that is, second order with y' absent, is discussed below. Note that the simple harmonic motion equation $y'' = -\omega^2 y$ is of this form, and that often a substitution can be found which will reduce a second-order equation containing y' to this form.

Two boundary conditions must be known for the solution of a second order differential equation, and it is here assumed that the values of y and y' are known for $x = 0$.

6.3.1 Step-by-step solution using central differences and a series start

This method has as its basis the use of the integration formula $\delta^2 y_0 = h^2(y_0'' + \frac{1}{12}\delta^2 y_0'' - \frac{1}{240}\delta^4 y_0'' + \ldots)$ the derivation of which may be found in more advanced texts (e.g. Hartree).

It is desirable to truncate this formula after second differences, so we require $(h^2 \delta^4 y_0'')/240 < \frac{1}{2}$ in terms of the last decimal place. $\delta^2 y_0$ may then be obtained from the equation $\delta^2 y_0 = h^2(y_0'' + 0 \cdot 083 \delta^2 y_0'')$.

The solution is begun using a power series to find a few values of y, and thus y'' by substitution in the original differential equation, and continued by estimation and correction of differences.

Worked example

Consider the solution of $y'' = -4y$ in the range $x = 0$ to $x = 0 \cdot 8$ with $y = \frac{1}{2}$ and $y' = 0$ when $x = 0$.

Series start

Assuming that $y = a_0 + a_1 x + a_2 x^2 + a_3 x^3 + a_4 x^4 + \ldots$ we have $a_0 = \frac{1}{2}$ as $y = \frac{1}{2}$ when $x = 0$.

Differentiating the series gives $y' = 1a_1 + 2a_2 x + 3a_3 x^2 + 4a_4 x^3 + \ldots$ but as $y' = 0$ when $x = 0$ we obtain $a_1 = 0$.

Differentiating the series again gives $y'' = 2a_2 + 6a_3 x + 12a_4 x^2 + \ldots$ Either by substituting these power series for y and y'' in the differential equation and comparing differences, or by continued differentiation and substitution we obtain the series:

$$y = \tfrac{1}{2} - x^2 + \tfrac{1}{3}x^4 - \tfrac{2}{45}x^6 + \ldots$$

This series is used to calculate the values of y for $x = -0\cdot2(0\cdot1)0\cdot2$, which are then substituted in the original differential equation to give corresponding values of y''. These values are tabulated and differenced as shown below

x	$y'' = -4y$				y			
$-0\cdot2$	$-1\cdot842\ 122$				$0\cdot460\ 530$			
		$-118\ 011$				$29\ 503$		
$-0\cdot1$	$-1\cdot960\ 133$		$78\ 144$		$0\cdot490\ 033$		$-19\ 536$	
		$-39\ 867$		$1\ 590$		$9\ 967$		-398
0	-2		$79\ 734$		$0\cdot5$		$-19\ 934$	
		$+39\ 867$		$-1\ 590$		$-9\ 967$		$+398$
$0\cdot1$	$-1\cdot960\ 133$		$78\ 144$		$0\cdot490\ 033$		$-19\ 536$	
		$118\ 011$				$-29\ 503$		
$0\cdot2$	$-1\cdot842\ 122$				$0\cdot460\ 530$			

Note: values $-3\ 180$ and 796 appear as the fourth differences centered at $x=0$.

The size of the interval being used may now be checked using the fourth difference in y''. It is desirable that $h^2(3180/240) < 0\cdot5$ which gives $h^2 < 0\cdot038$ or $h < 0\cdot19$, showing that the interval of $0\cdot1$ is satisfactory.

Estimation-correction process

On the right-hand side of the equation $\delta^2 y_0 = h^2(y_0'' + 0\cdot083\dot{3}\delta^2 y_0'')$ only $\delta^2 y_0''$ is unknown. This is estimated by referring to the known differences of y''. Here the next third difference of y'' may be estimated as -4600, giving an estimate of $73\ 544$ for $\delta^2 y_0''$. Thus we obtain

$$\delta^2 y_0 = (0\cdot1)^2(-1\cdot842\ 122 + 0\cdot083\ 333 \times 0\cdot073\ 544) = -0\cdot018\ 360$$

This working would be added to the tables of y and y'' as shown below.

x	$y'' = -4y$				y				
$0\cdot2$	$-1\cdot842\ 122$				$0\cdot460\ 530$				
		$-118\ 011$				$29\ 503$			
$0\cdot1$	$-1\cdot960\ 133$		$78\ 144$		$0\cdot490\ 033$		$-19\ 536$		
		$-39\ 867$		1590		$9\ 967$		-398	
0	-2		$79\ 734$		-3180	$0\cdot5$		$-19\ 934$		796
		$+39\ 867$		-1590		$-9\ 967$		$+398$	
$0\cdot1$	$-1\cdot960\ 133$		$78\ 144$		$0\cdot490\ 033$		$-19\ 536$		
		$118\ 011$		(-4600)		$-29\ 503$			
$0\cdot2$	$-1\cdot842\ 122$		$(73\ 544)$		$0\cdot460\ 530$		$-18\ 360$		

$\delta y_{\frac{1}{2}}$, and thus y_1, may now be found using this value of $\delta^2 y_0$. This is then substituted in the original differential equation to obtain y_1'' from which the estimated differences are corrected as shown below.

```
 x   | y'' = 4y
-0·2 | -1·842 122                                        0·460 530
     |            -118 011                                          29 503
-0·1 | -1·960 133            78 144                       0·490 033          -19 536
     |            -39 867            1590                            9 967          -19 536     -398
 0   | -2                    79 734          -3180        0·5                -19 934                796
     |            +39 867            -1590                          -9 967          -19 934     +398
 0·1 | -1·960 133            78 144          -3111        0·490 033          -19 536
     |            118 011            -4701                          -29 503
 0·2 | -1·842 122            73 443                       0·460 530          -18 360
     |            191 454                                          -47 863     -18 360
 0·3 | -1·650 668                                         0·412 667
```

The value of $\delta^2 y_0$ is checked using the new value of $\delta^2 y_0''$, 73 443, and further corrections made if necessary. The solution is now continued using the same process, being shown below as far as $x = 0.5$. The working has been carried out with six decimal places so that the values of y and y'' may be obtained correct to 4D. If values of y' are required then the table of values of y'' should be integrated using one of the methods of Chapter 4.

```
 x    | y'' = - 4y                                  |  y
- 0·2 | - 1·842 122                                 |  0·460 530
      |             - 118 011                       |             29 503
- 0·1 | - 1·960 133        78 144                   |  0·490 033    - 19 536
      |             - 39 867      1 590             |       9 967        - 398
 0    | - 2                79 734      - 3 180      |  0·5          - 19 934      796
      |             + 39 867  - 1 590               |     - 9 967        + 398
 0·1  | - 1·960 133        78 144      - 3 111      |  0·490 033    - 19 536      778
      |             118 011  - 4 701                |     - 29 503       1 176
 0·2  | - 1·842 122        73 443      - 2 936      |  0·460 530    - 18 360      732
      |             191 454  - 7 637                |     - 47 863       1 908
 0·3  | - 1·650 668        65 806      - 2 617      |  0·412 667    - 16 452      656
      |             257 260 - 10 254                |     - 64 315       2 564
 0·4  | - 1·393 408        55 552                   |  0·348 352    - 13 888
      |             312 812                         |     - 78 203
 0·5  | - 1·080 596                                 |  0·270 149
```

This example has been chosen to illustrate the method. Here an analytical solution is available which is $y = \frac{1}{2}\cos 2x$, giving at $x = 0.5$, $y = 0.270\ 151$ confirming the value obtained above. The method can be applied however to more difficult examples without any further complication.

6.3.2 Examples

1. The differential equation $y'' = 3xy$ is subject to the conditions $y = 1$, $y' = -1$ at $x = 0$. Obtain values of y to 3D over the range $x = 0(0.1)0.5$.

2. The differential equation $y'' = 2y^2$ is subject to the conditions $y = 0$, $y' = 1$ at $x = 0$. Obtain values of y to 3D over the range $x = 0(0 \cdot 1)0 \cdot 5$.

3. The differential equation $y'' + 3x^2y = \sin x$ is subject to the conditions $y = 1$, $y' = -1$ at $x = 0$. Obtain values of y to 4D over the range $x = 0(0 \cdot 1)0 \cdot 5$.

4. The differential equation $y'' - 4xy' + 2x^2y = 4$ is subject to the conditions $y = 0$, $y' = 1$ at $x = 0$. Show that the substitution $y = ze^{x^2}$ reduces this to the equation $z'' + (2 - 2x^2)z = 4e^{-x^2}$ subject to $z = 0$, $z' = 1$ at $x = 0$. By first finding values of z obtain a table of values of y to 4D over the range $x = 0(0 \cdot 05)0 \cdot 30$.

6.3.3 An approximate method of solution of $y'' = f(x, y)$

If second and higher differences are neglected in the integration formula used in § 6.3.1 we have $\delta^2 y_0 \simeq h^2 y_0''$. (The same result is obtained on truncating the differentiation formula $h^2 y_0'' = \delta^2 y_0 - \frac{1}{12}\delta^4 y_0 + \frac{1}{90}\delta^6 y_0 - \ldots$ after second differences.)

But $\delta^2 y_0 = \delta y_{\frac{1}{2}} - \delta y_{-\frac{1}{2}}$

$$= (y_1 - y_0) - (y_0 - y_{-1})$$

$$= y_1 - 2y_0 + y_{-1}$$

and as $y_0'' = f(x_0, y_0)$ from the original differential equation

then $y_1 - 2y_0 + y_{-1} \simeq h^2 f(x_0, y_0)$

so that $y_1 \simeq h^2 f(x_0, y_0) + 2y_0 - y_{-1}$.

This recurrence relation enables an approximate value of y to be obtained from two previous points. An approximate table of values of y can thus be obtained if either the values of y and y' are given at one point, or if two successive values of y are given. In the first case a second value of y will have to be found using a series.

Worked example

Consider again the solution of $y'' = -4y$ if $y = 0 \cdot 500$ at $x = 0$ and $0 \cdot 490$ at $x = 0 \cdot 1$.

Substituting $f(x, y) = -4y$ in the recurrence relation obtained above we have here $y_1 \simeq -4h^2 y_0 + 2y_0 - y_{-1}$.

Putting $h = 0 \cdot 1$, $y_0 = 0 \cdot 490$ and $y_{-1} = 0 \cdot 500$ gives $y_1 = 0 \cdot 460$ to 3D at $x = 0 \cdot 2$. Then with $y_0 = 0 \cdot 460$ and $y_{-1} = 0 \cdot 490$ we have at $x = 0 \cdot 3$, $y = 0 \cdot 412$ to 3D. Continuing with this process, and working to 3D, the following table of values of y is obtained:

x	0	0·1	0·2	0·3	0·4	0·5	0·6	0·7	0·8
y	0·500	0·490	0·460	0·412	0·348	0·270	0·181	0·085	− 0·014

This table may be extended backwards by rewriting the recurrence relation to give y_{-1} in terms of y_0 and y_1. The method may be developed so as to take account of the neglected differences.

6.3.4 Examples

1. The differential equation $y'' = 3xy$ is subject to the conditions $y = 1$ at $x = 0$ and $y = 0.9$ at $x = 0.1$. Obtain approximate values of y over the range $x = 0(0.1)0.5$.

2. The differential equation $y'' = 2y^2$ is subject to the conditions $y = 0$ at $x = 0$, $y = 0.1$ at $x = 0.1$. Obtain approximate values of y over the range $x = 0(0.1)0.5$.

3. The differential equation $y'' + 3x^2y = \sin x$ is subject to the conditions $y = 1$ at $x = 0$ and $y = 0.9$ at $x = 0.1$. Obtain approximate values of y over the range $x = 0(0.1)0.5$.

4. The differential equation $\ddot{x} = -0.25x$ is subject to the conditions $x = 1$, $\dot{x} = 0$ at $t = 0$. Obtain approximate values of x over the range $t = 0(0.1)1.0$ and compare your results with those found from the analytical solution.

7

Curve Fitting by the Method of Least Squares

7.1 INTRODUCTION

It is frequently desirable to obtain a relation between two variables for which a set of experimental results has been obtained. For example when a new machine has been designed and constructed its performance must be tested. Certain measurements will be made and the results may appear as follows:

x	20	30	40	50	100	150	200
y	10	20	35	50	150	300	500

Here the variable x could be the engine speed and y the fuel consumption, or x could be the load applied and y the resulting stress in a particular component, these quantities being measured in appropriate units.

The manufacturer needs to know the relation between the variables so that he can find and use intermediate values, or estimate values lying just outside the range of the tests, possibly with the design of further tests in mind. The functional relation between the variables may be known from theoretical analysis. In this case a method is required to obtain the most probable values of the constants which appear in the formula. The method discussed in Vol. I, Chapter 5, can be used when it is known that the given set of points lie *exactly* on some polynomial. On the other hand the functional relation may be completely unknown, and in this case we need to find some empirical formula which will approximate to the unknown function, and represent it accurately enough over a limited range of values of the variables. The interpolation formulae introduced in Chapter 1 are polynomial approximations to functions, but they are only applicable over a small range of values and it is assumed that the data is subject only to rounding off errors.

The method described below is applicable to a set of points $(x_1, y_1), (x_2, y_2) \ldots$ (x_m, y_m) which are obtained experimentally. These values may be subject to errors (greater than any rounding off errors) the magnitude of which may be unknown, or for which only the probable maximum value is available. In most experimental situations the values of x are assumed to be exact and the only errors are in the values of y. The other possibilities—of errors only in the values of x or of errors in both the values of x and y—are not discussed here (see any text book of Statistics referring to regression lines).

Interpreting the problem graphically, if the points $(x_1, y_1), (x_2, y_2) \ldots (x_m, y_m)$ are plotted then it is desired to draw a curve which passes as near as possible to all the points and which smooths out experimental errors. Any numerical method used to obtain such a curve depends for its success on the assumption that the errors involved are random, and that on average they tend to cancel out.

There is no curve which is *the* best fitting curve, as the one obtained depends on the criterion used. For example, consider the problem of fitting a straight line.

7.2   FITTING A STRAIGHT LINE

In fig. 7.1 the distance of the point P_r from the given line may be defined in any of the following ways:

(i) P_rN, the perpendicular distance of the point from the line.

(ii) P_rR, the distance parallel to the x-axis.

(iii) P_rQ, the distance parallel to the y-axis.

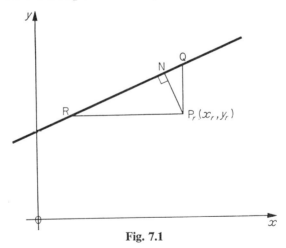

Fig. 7.1

But the distances of all the points from the line have to be considered and if the error in P_r, defined in one of the above ways, is e_r, then it would be possible to choose the line to satisfy any of the following conditions:

(i) The algebraic sum of the errors is to be zero, i.e. $\Sigma e_r = 0$.

(ii) The sum of the magnitudes of the errors is to be a minimum, i.e. $\Sigma |e_r|$ is a minimum.

(iii) The sum of the squares of the errors is to be a minimum, i.e. $\Sigma(e_r)^2$ is a minimum, etc.

We will apply the method of *Least Squares*, which is that most frequently used, and assume that the errors are defined as the distances parallel to the y-axis, (such as P_rQ in figure 7.1). The curve is chosen so that the sum of the squares of these errors is a minimum, as shown in the following sections.

7.2.1 Fitting a straight line

Suppose the equation of the line is $y = a_0 + a_1x$.

Note that a set of values of x and y is given, and that it is the values of a_0 and a_1 which are the unknowns and which are subsequently treated as variables. Here $e_r = a_0 + a_1x_r - y_r$ and we require Σe_r^2 to be a minimum, i.e. $\Sigma(a_0 + a_1x_r - y_r)^2$ to be a minimum.

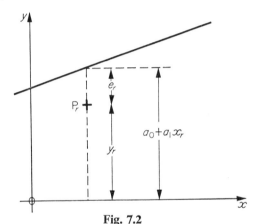

Fig. 7.2

As this expression contains two variables, a_0 and a_1 it is necessary to differentiate partially with respect to each of these variables in turn, putting these derivatives equal to zero to obtain the conditions for a minimum value of the expression. Owing to the nature of $\Sigma(a_0 + a_1x_r - y_r)^2$ it can have only one minimum value. (See note on maximum and minimum values in Vol I, §1.2.3).

Here differentiating $\sum\limits_r (a_0 + a_1x_r - y_r)^2$ firstly with respect to a_0, treating a_1 as a constant we require:

$$\frac{\partial}{\partial a_0} \sum_r (a_0 + a_1x_r - y_r)^2 = 0$$

i.e.
$$\sum_r 2(a_0 + a_1x_r - y_r) = 0$$

giving
$$2a_0m + 2a_1 \sum_r x_r - 2\sum_r y_r = 0$$

so that
$$a_0m + a_1 \sum_r x_r = \sum y_r \qquad (1)$$

Secondly, differentiating with respect to a_1 and treating a_0 as a constant we require:

$$\frac{\partial}{\partial a_1} \sum_r (a_0 + a_1 x_r - y_r)^2 = 0$$

i.e.

$$\sum_r 2(a_0 + a_1 x_r - y_r)x_r = 0$$

giving

$$2a_0 \sum_r x_r + 2a_1 \sum_r x_r^2 - 2 \sum_r x_r y_r = 0$$

so that

$$a_0 \sum_r x_r + a_1 \sum_r x_r^2 = \sum_r x_r y_r \qquad (2)$$

So we obtain a pair of simultaneous equations in a_0 and a_1 namely

$$a_0 m + a_1 \sum_r x_r = \sum_r y_r \qquad (1)$$

$$a_0 \sum_r x_r + a_1 \sum_r x_r^2 = \sum_r x_r y_r \qquad (2)$$

the solution of which gives the required values of a_0 and a_1. These equations are called the normal equations.

Worked example

To fit a straight line to the following data:

x	0·5	1·0	1·5	2·0	2·4	2·6	2·8	3·0
y	0·22	0·24	0·27	0·33	0·35	0·35	0·38	0·39

The coefficients of the normal equations: $m, \sum_r x_r, \sum_r x_r^2, \sum_r y_r$ and $\sum_r x_r y_r$, are calculated using the following table for the working and checks. Each row is evaluated in turn as follows:

Set x_r in the S.R. and multiply by x_r to obtain x_r^2.

Clear the C.R. and accumulator and multiply by y_r to obtain $x_r y_r$.

Clear the C.R. and accumulator and add 1 to x_r *in the S.R.*

Add into the accumulator then clear the C.R. and S.R.

Set y_r in the S.R. and add into the accumulator to give $s_r = 1 + x_r + y_r$.

Back transfer and multiply by x_r to give $x_r s_r$.

Each column is then totalled to give $m, \sum_r x_r, \sum_r y_r, \sum_r x_r^2, \sum_r x_r y_r, \sum_r s_r$ and $\sum_r x_r s_r$ respectively, checking that $\sum_r s_r = m + \sum_r x_r + \sum_r y_r$ and that

$$\sum_r x_r s_r = \sum_r x_r + \sum_r x_r^2 + \sum_r x_r y_r.$$

1	x_r	y_r	x_r^2	$x_r y_r$	s_r	$x_r s_r$
1	0·5	0·22	0·25	0·110	1·72	0·860
1	1·0	0·24	1·00	0·240	2·24	2·240
1	1·5	0·27	2·25	0·405	2·77	4·155
1	2·0	0·33	4·00	0·660	3·33	6·660
1	2·4	0·35	5·76	0·840	3·75	9·000
1	2·6	0·35	6·76	0·910	3·95	10·270
1	2·8	0·38	7·84	1·064	4·18	11·704
1	3·0	0·39	9·00	1·170	4·39	13·170
8	15·8	2·53	36·86	5·399	26·33	58·059

Here the normal equations are:

$$8a_0 + 15\cdot8a_1 = 2\cdot53$$
$$15\cdot8a_0 + 36\cdot86a_1 = 5\cdot399$$

and the values of a_0 and a_1 can be found by one of the methods of Vol I, Chapter 7, to give $a_0 = 0\cdot176$ and $a_1 = 0\cdot071$ correct to 3D.

i.e. The required straight line is:

$$y = 0\cdot176 + 0\cdot071x.$$

The rounded values of y obtained from this equation at the given values of x are shown in the following table, below the given values of y.

x_r	0·5	1·0	1·5	2·0	2·4	2·6	2·8	3·0
y_r	0·22	0·24	0·27	0·33	0·35	0·35	0·38	0·39
$0\cdot176 + 0\cdot071x_r$	0·21	0·25	0·28	0·32	0·35	0·36	0·38	0·39

7.2.2 Alternative method of calculating the coefficients

The above method of calculating the coefficients for the normal equations can be extended to find the normal equations for other functions, but in the case of the straight line the following alternative method may be used. This is rather quicker than the above method but it has the disadvantage that it does not incorporate any check.

The coefficients are found in pairs without intermediate writing down as follows:

(a) $\sum_r x_r$ and $\sum_r y_r$

Mark off both the S.R. and the accumulator to 2D and to 7D.

Set the x values so that the units position is position 8.

Set the y values so that the units position is position 3.

For example, x_1 and y_1 will be set as $0 \cdot 500\ 00 \cdot 22$. Add this into the accumulator.

Set x_2 and y_2 as $1 \cdot 000\ 00 \cdot 24$, add and continue until all the x and y values have been added.

The accumulator now reads $15 \cdot 800\ 02 \cdot 53$ giving $\sum_r x_r = 15 \cdot 8$ and $\sum_r y_r = 2 \cdot 53$.

(b) $\sum_r x_r^2$ and $\sum_r x_r y_r$

With the decimal markers in the same positions in the S.R. mark off the C.R. to 1D and the accumulator to 3D and 8D. Set x_1 and y_1 as before but multiply by x_1 to give x_1^2 and $x_1 y_1$ in the accumulator. Repeat with the other pairs of values of x and y, the first few steps being as follows:

	C.R.	S.R.	Acc.
$\left\{\begin{array}{l}\text{Set } x_1 = 0 \cdot 5,\ y_1 = 0 \cdot 22 \\ \text{Mult. by } x_1 = 0 \cdot 5\end{array}\right.$	0 0·5	0·500 00·22 0·500 00·22	0 0·250 00·110
$\left\{\begin{array}{l}\text{Clear S.R. \& C.R. Set} \\ \quad x_2 = 1 \cdot 0,\ y_2 = 0 \cdot 24 \\ \text{Mult. by } x_2 = 1 \cdot 0\end{array}\right.$	0 1·0	1·000 00·24 1·000 00·24	0·250 00·110 1·250 00·350
$\left\{\begin{array}{l}\text{Clear S.R. \& C.R. Set} \\ \quad x_3 = 1 \cdot 5,\ y_3 = 0 \cdot 27 \\ \text{Mult. by } x_3 = 1 \cdot 5 \text{ etc.} \\ \qquad \text{etc.}\end{array}\right.$	0 1·5	1·500 00·27 1·500 00·27	1·250 00·350 3·500 00·755

After the final multiplication by $x_8 = 3 \cdot 0$ the accumulator contains $36 \cdot 860\ 05 \cdot 399$ giving $\sum_r x_r^2 = 36 \cdot 86$ and $\sum_r x_r y_r = 5 \cdot 399$.

Note.

The S.R. and accumulator must have sufficient capacity to contain the various totals without overflow. This must also be taken into account when the positions of the two decimal markers in the S.R. are chosen.

7.2.3 Change of origin

In some cases it may be possible to simplify the working by changing the origin of either or both of the variables. For example, if a curve of the form $y = f(x)$ is required to fit the following data:

x	14·2	14·4	14·6	14·8	15·0
y	53·1	57·6	61·3	65·9	69·4

the working is easier if new variables $X = x - 14$ and $Y = y - 50$ are used. i.e. Find a curve of the form $Y = F(X)$ to fit the data.

X	0·2	0·4	0·6	0·8	1·0
Y	3·1	7·6	11·3	15·9	19·4

and substitute $X = x - 14$ and $Y = y - 50$ in this to obtain $y = f(x)$.

It can be shown that the straight line obtained by the method of least squares passes through the point $(\bar{x}, \bar{y})$, where x and y are the arithmetic means of the x and y values respectively. If the origin is changed to $(\bar{x}, \bar{y})$ by the transformation $X = x - \bar{x}$, $Y = y - \bar{y}$, then the equation of the required straight line reduces to the form $Y = a_1 X$ and the normal equations are simply $ma_0 = 0$, giving $a_0 = 0$, and $a_1 \sum_r X_r^2 = \sum_r X_r Y_r$ giving the value of a_1.

In the previous worked example, in § 7.2.1, $\bar{x} = 15\cdot8 \div 8 = 1\cdot975$ and $\bar{y} = 2\cdot53 \div 8 = 0\cdot31625$. If new variables $X = x - 1\cdot975$ and $Y = y - 0\cdot316\,25$ are used then the straight line obtained is $Y = 0\cdot071X$ which can then be easily transformed into the required form $y = 0\cdot176 + 0\cdot071x$.

Although the normal equations are particularly simple when this method is used, more working is needed to obtain them, so that it is not always the quickest method.

7.2.4 The straight line not necessarily suitable

The straight line which is fitted to a set of points by the method of least squares will be the best fitting *straight line* given by this method, but it may not fit the points satisfactorily, and a quadratic or higher polynomial may be preferable. For example, given the set of points in figure 7.2.4 the method of least squares may yield the straight line shown, but this clearly does not fit the points as satisfactorily as the curve.

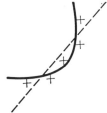

Fig. 7.2.4

Some thought should first be given to the most suitable polynomial to use for a particular set of points, bearing in mind the subsequent application of the result. However, having fitted a curve, guidance in deciding if it is a satisfactory fit can be given by further comparison of the data and the calculated values. (see more advanced texts for details.).

7.2.5 Some functions are reducible to a linear relationship, by suitable changes in the variables, and may be rewritten so that the representation is by straight lines. Four examples follow.

1. $y = ax^n$

Take the logarithm of each side. (The base is not important and depends on the tables available).

Thus $\quad\qquad\qquad\qquad \log y = \log (ax^n)$

$$= \log a + \log x^n$$

so that $\qquad\qquad\qquad \log y = \log a + n \log x$

Putting $\qquad\qquad\qquad \log y = Y \text{ and } \log x = X \text{ gives}$

$$Y = \log a + nX$$

which is the equation of a straight line.

Thus by tabulating $\log y$ and $\log x$ and using these values to obtain the normal equations we have $\log a = a_0$ giving the value of a, and $n = a_1$.

2. $y = \log_e (a + bx)$

Rewrite this as $e^y = a + bx$ and use e^{y_r} instead of y_r to obtain the normal equations.

3. $y = a + b \sin x$

Use $\sin x$ instead of x to obtain the normal equations.

4. $y = ax + bx^2$

Rewrite this as $(y/x) = a + bx$ and use (y_r/x_r) instead of y_r to obtain the normal equations.

7.2.6   Examples

1. Fit a straight line to the following data obtaining the coefficients to 3D.

x	0	1	2	3	4	5	6	7
y	2·42	2·55	2·96	3·08	3·37	3·60	3·79	4·04

2. Fit a straight line to the following data obtaining the coefficients to 4S.

x	0	5	10	15	20	25	30
y	21·9	22·3	22·8	23·1	23·7	24·3	34·9

3. Fit a straight line to the following data obtaining the coefficients to 4S.

x	12	14	16	18	20	22
y	8·85	8·31	7·67	7·01	6·39	5·68

4. Fit a straight line to the following data obtaining the coefficients to 3D.

x	1	2	3	4	6	8	10	12
y	0·41	0·37	0·34	0·30	0·25	0·18	0·10	0·05

5. Find values of a and c correct to 2D in order to fit a curve of the form $y = ae^{cx}$ to the following data.

x	0	0·01	0·02	0·03	0·04	0·05	0·06
y	2·02	2·07	2·12	2·18	2·25	2·34	2·41

6. By the method of Least Squares fit a straight line to the following data, assuming that the values of x are exact.

x	1·9	3·5	4·7	5·7	8·8	15·7	22·5	32·0
y	1·01	1·45	1·76	2·05	2·90	4·76	6·64	9·20

(A.E.B. 1960)

7. Find, by the method of least squares, a formula of the type $y = ax^n$ which will fit the following data. It may be assumed that the values of x are exact.

x	1·0	1·15	1·4	1·43	1·6	2·0	2·35	2·7
y	4·33	4·58	4·98	5·06	5·28	5·80	6·24	6·49

(A.E.B. 1961)

7.3 FITTING A POLYNOMIAL

Let the equation of the required polynomial be

$$y = a_0 + a_1x + a_2x^2 + \ldots + a_nx^n$$

and the given points (x_r, y_r), $r = 1, 2, \ldots m$, $n < m$, where again the x values are assumed to be exact and only the y values in error. Then the error at the rth point is $e_r = a_0 + a_1 x_r + a_2 x_r^2 + \ldots + a_n x_r^n - y_r$ and we require the values of $a_0, a_1, \ldots, a_n$ which give a minimum value of $\sum_r e_r^2$.

Using partial differentiation for each of the variables $a_0, a_1, \ldots, a_n$ in turn, as for the straight line case, $n + 1$ normal equations are obtained which are:

$$a_0 m + a_1 \sum_{r=1}^{m} x_r + a_2 \sum_{r=1}^{m} x_r^2 + \ldots + a_n \sum_{r=1}^{m} x_r^n = \sum_{r=1}^{m} y_r$$

$$a_0 \sum_{r=1}^{m} x_r + a_1 \sum_{r=1}^{m} x_r^2 + a_2 \sum_{r=1}^{m} x_r^3 + \ldots + a_n \sum_{r=1}^{m} x_r^{n+1} = \sum_{r=1}^{m} x_r y_r$$

$$a_0 \sum_{r=1}^{m} x_r^2 + a_1 \sum_{r=1}^{m} x_r^3 + a_2 \sum_{r=1}^{m} x_r^4 + \ldots + a_n \sum_{r=1}^{m} x_r^{n+2} = \sum_{r=1}^{m} x_r^2 y_r$$

$$a_0 \sum_{r=1}^{m} x_r^n + a_1 \sum_{r=1}^{m} x_r^{n+1} + a_2 \sum_{r=1}^{m} x_r^{n+2} + \ldots + a_n \sum_{r=1}^{m} x_r^{2n} = \sum_{r=1}^{m} x_r^n y_r$$

The coefficients of these normal equations may be calculated by an extension of the method used in the straight line case.

Worked example

Fit a curve of the form $y = a_0 + a_1 x + a_2 x^2$ to the following data:

x	1	2	3	4	5	6	7	8	9
y	2·1	2·3	3·9	4·4	4·6	4·8	4·6	4·2	3·4

The normal equations to be found are:

$$a_0 m + a_1 \sum_r x_r + a_2 \sum_r x_r^2 = \sum_r y_r$$

$$a_0 \sum_r x_r + a_1 \sum_r x_r^2 + a_2 \sum_r x_r^3 = \sum_r x_r y_r$$

$$a_0 \sum_r x_r^2 + a_1 \sum_r x_r^3 + a_2 \sum_r x_r^4 = \sum_r x_r^2 y_r$$

It is necessary to calculate the sums of the powers of the x values up to the fourth power of x, and the sums of the products $x_r y_r$ and $x_r^2 y_r$. The check sum s_r is now $s_r = 1 + x_r + x_r^2 + y_r$ and two other check columns $x_r s_r$ and $x_r^2 s_r$ are required. (In general $s_r = 1 + x_r + x_r^2 + \ldots + x_r^n + y_r$ and further check columns up to $x_r^n s_r$ are required. There is one check column for each equation, and their sums are the sums of the coefficients of the corresponding equations). The checks to be made are that

$$\sum_r s_r = m + \sum_r x_r + \sum_r x_r^2 + \sum_r y_r,$$

$$\sum_r x_r s_r = \sum_r x_r + \sum_r x_r^2 + \sum_r x_r^3 + \sum_r x_r y_r$$

$$\text{and} \quad \sum_r x_r^2 s_r = \sum_r x_r^2 + \sum_r x_r^3 + \sum_r x_r^4 + \sum_r x_r^2 y_r.$$

The working is simplified by changing the origin of the x values, using $X_r = x_r - 5$ and is shown below.

When such a change of origin results in simple integer values of x it is usually quicker to complete the columns of powers of x before calculating any other parts of the table. In general however, the table should be completed row by row using the following operations on the calculating machine:

Set X_r; multiply by X_r to give X_r^2, back transfer and multiply by X_r to give X_r^3, back transfer and multiply by X_r to give X_r^4.

Set y_r; multiply by X_r to give $X_r y_r$, back transfer and multiply by X_r to give $X_r^2 y_r$.

Evaluate s_r; back transfer and multiply by X_r to give $X_r s_r$, back transfer and multiply by X_r to give $X_r^2 s_r$.

1	X_r	y_r	X_r^2	$X_{r_2}^3$	X_r^4	$X_r y_r$	$X_r^2 y_r$	s_r	$X_r s_r$	$X_r^2 s_r$
1	-4	2·1	16	-64	256	$-8·4$	33·6	15·1	$-60·4$	241·6
1	-3	3·3	9	-27	81	$-9·9$	29·7	10·3	$-30·9$	92·7
1	-2	3·9	4	-8	16	$-7·8$	15·6	6·9	$-13·8$	27·6
1	-1	4·4	1	-1	1	$-4·4$	4·4	5·4	$-5·4$	5·4
1	0	4·6	0	0	0	0	0	5·6	0	0
1	1	4·8	1	1	1	4·8	4·8	7·8	7·8	7·8
1	2	4·6	4	8	16	9·2	18·4	11·6	23·2	46·4
1	3	4·2	9	27	81	12·6	37·8	17·2	51·6	154·8
1	4	3·4	16	64	256	13·6	54·4	24·4	97·6	390·4
9	0	35·3	60	0	708	9·7	198·7	104·3	69·7	966·7

Here the normal equations are:

$$9a_0 \qquad\quad + 60a_2 = 35·3$$
$$60a_1 \qquad\qquad = 9·7$$
$$60a_0 \qquad\quad + 708a_2 = 198·7$$

which also are simpler than those which would have been obtained if no change of origin had been made. The solution of these to 4D is $a_0 = 4·7151$, $a_1 = 0·1617$ and $a_2 = -0·1189$, giving the equation in terms of X:

$$y = 4·7151 + 0·1617\, X - 0·1189\, X^2$$

Substituting back gives:

$$y = 4·7151 + 0·1617\, (x - 5) - 0·1189\, (x - 5)^2$$

which simplifies to

$$y = 0·9433 + 1·3507x - 0·1189x^2$$

so that the required equation, to 3D, is:

$$y = 0.943 + 1.351x - 0.119x^2.$$

The values of y given by this equation on rounding off to 2D are compared below with the original values.

x_r	1	2	3	4	5	6	7	8	9
y_r	2·1	3·3	3·9	4·4	4·6	4·8	4·6	4·2	3·4
Calculated y_r	2·18	3·17	3·93	4·44	4·72	4·77	4·57	4·14	3·46

The coefficients of the normal equations for polynomials of a higher degree may be obtained in a similar way by including further columns in the working

7.3.1   Examples

1. Fit a quadratic to the following data, obtaining the coefficients correct to 4S.

x	0	1	2	3	4	5	6
y	0	2	5	9	15	22	29

2. Fit a quadratic to the following data, obtaining the coefficients correct to 4S.

x	0·0	0·1	0·2	0·3	0·4	0·5	0·6
y	50·6	52·5	53·6	54·3	54·8	55·2	55·4

3.

x	0	2	4	6	8	10	12	14	16
y	101	104	107	107	108	110	109	110	108

In order to fit an equation of the form $y = a + bx + cx^2$ to the above data, a false origin ($y = 100$) was used.
Verify that the following are suitable normal equations and solve them to determine the quadratic curve of best fit.

$$9a + \qquad 72b + \qquad 816c = \qquad 64$$
$$72a + \qquad 816b + \quad 10\,368c = \quad 618$$
$$816a + 10\,368b + 140\,352c = 7\,196 \qquad \text{(A.E.B. 1963)}$$

(Note that in fact it is easier to change the origin of the values of x as well as that of the values of y.)

4. The heights of a projectile in feet at intervals of one second are given in the following table:

t	0	1	2	3	4	5	6	7	8
h	0	7	13	17	21	23	25	26	27

Obtain the equation of the parabolic path, giving the coefficients to 3S.

5. Fit a quadratic to the following data obtaining the coefficients to 2D.

x	0·2	0·4	0·6	0·8	1·0	1·2
y	3·8	1·7	0·9	1·0	1·9	3·5

6. Fit a quadratic to the following data obtaining the coefficients to the nearest integer.

x	0	1	2	3	4	5	6	7
y	100	99	96	86	72	56	35	10

7.4 FITTING OTHER FUNCTIONS

Functions of the form $y = a_0 f_0(x) + a_1 f_1(x) + \ldots + a_n f_n(x)$ (e.g. $y = ae^x + be^{-x}$) may be fitted to appropriate data by the same method as that for polynomials, the number of constants determining the number of normal equations.

e.g. Consider functions of the form $y = af(x) + bg(x)$.

Here
$$e_r = af(x_r) + bg(x_r) - y_r$$

so that
$$\sum_r e_r^2 = \sum_r (af(x_r) + bg(x_r) - y_r)^2$$

For $\sum_r e_r^2$ to be a minimum it is necessary that $(\partial/\partial a)\sum_r e_r^2 = 0$ and $(\partial/\partial b)\sum_r e_r^2 = 0$.

i.e.
$$\frac{\partial}{\partial a}\sum_r [af(x_r) + bg(x_r) - y_r]^2 = 0$$

giving
$$2\sum_r [af(x_r) + bg(x_r) - y_r]f(x_r) = 0$$

so that
$$a\sum_r f(x_r) \cdot f(x_r) + b\sum_r f(x_r) \cdot g(x_r) = \sum_r y_r f(x_r)$$

and
$$\frac{\partial}{\partial b}\sum_r [af(x_r) + bg(x_r) - y_r]^2 = 0$$

giving
$$2\sum_r [af(x_r) + bg(x_r) - y_r)] g(x_r) = 0$$

so that $a\sum_r f(x_r) g(x_r) + b\sum_r g(x_r) g(x_r) = \sum_r y_r g(x_r)$

i.e. The normal equations are:

$$a\sum_r f(x_r).f(x_r) + b\sum_r f(x_r).g(x_r) = \sum_r y_r.f(x_r)$$

$$a\sum_r f(x_r).g(x_r) + b\sum_r g(x_r).g(x_r) = \sum_r y_r.g(x_r)$$

The values of the coefficients may be obtained using a similar layout as that for polynomials. Suitable columns would be x_r, y_r, $f(x_r).f(x_r)$, $f(x_r).g(x_r)$, $y_r f(x_r)$, $g(x_r).g(x_r)$, $y_r g(x_r)$ with two check columns $s_r = f(x_r)[f(x_r) + g(x_r) + y_r]$ and $c_r = g(x_r)[f(x_r) + g(x_r) + y_r]$.

7.4.1 Examples

1. Find values of a and b correct to 2D to fit a curve of the form $y = ae^x + be^{-x}$ to the following data:

x	0	0·5	1·0	1·5	2·0	2·5
y	5·02	5·21	6·49	9·54	16·02	24·53

2. Use the method of least squares to fit the curve

$$y = a \sin x + b \cos x$$

to the following data and hence suggest a value of y for $x = 0.5$:

x	0·0	0·1	0·3	0·4	0·7	0·9	1·0
y	1·56	1·77	2·19	2·37	2·75	2·88	2·97

(A.E.B. 1964)

8
Summation of Series with Slow Convergence

8.1 INTRODUCTION

In Vol I, §2.5, we discussed the evaluation of terms of a series, and hence its sum, for some series in which the convergence is fairly rapid. In this chapter we outline some methods for evaluating series for which the convergence is slow. These all rely on finding some means of transforming the series in order to improve the convergence.

Note that it is usually preferable to solve the given problem without using a series at all if this is possible.

8.2 EULER'S METHOD FOR ALTERNATING SERIES

For alternating series of the form $S = u_0 - u_1 + u_2 - u_3 + u_4 - \ldots$ where each $u_r > 0$ and also $u_r > u_{r+1}$ and $u_r \to 0$ as $r \to \infty$.

If the terms u_r are tabulated as a function of r and differenced then $u_r = E^r u_0$ and thus

$$S = (1 - E + E^2 - E^3 + \ldots)\, u_0$$

$$= (1 + E)^{-1}\, u_0 \quad \text{by the binomial theorem}$$

$$= (2 + \Delta)^{-1}\, u_0 \text{ since } E = 1 + \Delta$$

$$= \tfrac{1}{2}(1 + \tfrac{1}{2}\Delta)^{-1} u_0$$

$$= \tfrac{1}{2}(1 - \tfrac{1}{2}\Delta + \tfrac{1}{4}\Delta^2 - \tfrac{1}{8}\Delta^3 + \ldots) u_0 \quad \text{by the binomial theorem}$$

and hence

$$S = \tfrac{1}{2}(u_0 - \tfrac{1}{2}\Delta u_0 + \tfrac{1}{4}\Delta^2 u_0 - \tfrac{1}{8}\Delta^3 u_0 + \ldots)$$

If the series is converging slowly then the successive terms $u_0, u_1, u_2 \ldots$ wil, not differ much in value, so that the differences $\Delta u_0, \Delta^2 u_0, \Delta^3 u_0 \ldots$ will be small and hence this new series will converge more rapidly.

Worked example:

$$S = \tfrac{1}{2} - \tfrac{2}{3}x + \tfrac{3}{4}x^2 - \tfrac{4}{5}x^3 + \ldots \text{ for } x = 0\cdot 9$$

The first eleven terms are here evaluated to 5D and tabulated and differenced as below. The number of terms it is necessary to evaluate depends on the accuracy required and the magnitude of the differences.

```
 r     u_r

 0    0·500 00
              10 000
 1    0·600 00              -9250
               750                 6070
 2    0·607 50             -3180          -4105
             -2430                 1965          2938
 3    0·583 20             -1215          -1167          -2202
             -3645                  798           736           1706
 4    0·546 75              -417           -431           -496          -1348
             -4062                  367           240           358
 5    0·506 13               -50           -191           -138          -282
             -4112                  176           102            76
 6    0·465 01               126            -89            -62           -30
             -3986                   87            40            46
 7    0·425 15               213            -49            -16
             -3773                   38            24
 8    0·387 42               251            -25
             -3522                   13
 9    0·352 20               264
             -3258
10    0·319 62
```

The transformed series $S = \frac{1}{2}(u_0 - \frac{1}{2}\Delta u_0 + \frac{1}{4}\Delta^2 u_0 - \frac{1}{8}\Delta^3 u_0 + \ldots)$ is first used to sum the series beginning at the second term, and then again beginning at the third term as a check. The differences used in each case are underlined in the above table.

(a) Transforming the series from the second term:

$$S = 0 \cdot 5 - \tfrac{1}{2}[0 \cdot 6 - \tfrac{1}{2}(0 \cdot 007\ 50) + \tfrac{1}{4}(- 0 \cdot 031\ 80) - \tfrac{1}{8}(0 \cdot 019\ 65)$$
$$+ \tfrac{1}{16}(- 0 \cdot 0116\ 7) - \tfrac{1}{32}(0 \cdot 007\ 36) + \tfrac{1}{64}(- 0 \cdot 004\ 96)$$
$$- \tfrac{1}{128}(0 \cdot 003\ 58) + \tfrac{1}{256}(- 0 \cdot 002\ 82) \ldots]$$
$$= 0 \cdot 5 - \tfrac{1}{2}[0 \cdot 6 - (0 \cdot 003\ 75 + 0 \cdot 007\ 95 + 0 \cdot 002\ 46$$
$$+ 0 \cdot 000\ 73 + 0 \cdot 000\ 23 + 0 \cdot 000\ 08 + 0 \cdot 000\ 03 + 0 \cdot 000\ 01)]$$
$$= 0 \cdot 5 - \tfrac{1}{2}[0 \cdot 584\ 76]$$

Thus $S = 0 \cdot 207\ 62$

(b) Transforming the series from the third term:

$$S = 0 \cdot 5 - 0 \cdot 6 + \tfrac{1}{2}[0 \cdot 607\ 50 - \tfrac{1}{2}(- 0 \cdot 024\ 30) + \tfrac{1}{4}(- 0 \cdot 012\ 15)$$
$$- \tfrac{1}{8}(0 \cdot 007\ 98) + \tfrac{1}{16}(- 0 \cdot 004\ 31) - \tfrac{1}{32}(0 \cdot 002\ 40)$$
$$+ \tfrac{1}{64}(- 0 \cdot 001\ 38) - \tfrac{1}{128}(0 \cdot 000\ 76) \ldots]$$
$$= - 0 \cdot 1 + \tfrac{1}{2}[0 \cdot 607\ 50 + 0 \cdot 012\ 15 - (0 \cdot 003\ 04 + 0 \cdot 001\ 00$$
$$+ 0 \cdot 000\ 27 + 0 \cdot 000\ 08 + 0 \cdot 000\ 01)]$$
$$= - 0 \cdot 1 + \tfrac{1}{2}[0 \cdot 615\ 25]$$

Thus $S = 0.207\,63$

Hence correct to 4D $S = 0.2076$

8.3 EULER–MACLAURIN INTEGRATION FORMULA METHOD FOR NON-ALTERNATING SERIES

This method is applicable to series $\overset{n}{\underset{r=0}{\Sigma}} f(r)$, either convergent or divergent, in which the terms do not alternate in sign, provided that $f(r)$ can be integrated and differentiated with respect to r.

The Euler–Maclaurin integration formula (*I.A.T.*, page 67) with $h = 1$ gives:

$$\int_0^n f(r)\,dr = \tfrac{1}{2}f_0 + f_1 + f_2 + \ldots + f_{n-1} + \tfrac{1}{2}f_n$$
$$- \tfrac{1}{12}(f_n' - f_0') + \tfrac{1}{720}(f_n''' - f_0''') - \ldots$$

which can be rearranged as:

$$f_0 + f_1 + f_2 + \ldots + f_n = \int_0^n f(r)\,dr + \tfrac{1}{2}(f_0 + f_n) +$$
$$\tfrac{1}{12}(f_n' - f_0') - \tfrac{1}{720}(f_n''' - f_0''') + \ldots$$

The method is illustrated by a simple example which is such that it may also be computed easily by alternative methods.

Worked example

Find the sum of the first 21 terms of the series $1.2 + 2.3 + 3.4 + \ldots$

The whole series is first transformed and evaluated, and then transformed and evaluated beginning at the third term as a check.

(i) The sum required is $\overset{20}{\underset{r=0}{\Sigma}} f(r)$, where here $f(r) = (r + 1)(r + 2)$

$$= r^2 + 3r + 2.$$

so that $\displaystyle\int_0^{20} f(r)\,dr = \left[\frac{r^3}{3} + \frac{3r^2}{2} + 2r\right]_0^{20}$

which gives $\int_0^{20} f(r)\,dr = 3{,}306\tfrac{2}{3}$

Also since $f(r) = r^2 + 3r + 2$, $f_0 = 2$ and $f_{20} = 462$.

$f'(r) = 2r + 3$ and so $f_0' = 3$ and $f_{20}' = 43$. $f''(r) = 2$ so that $f'''(r)$ and higher derivatives are all zero.

Substituting in the above formula thus gives

$$\overset{20}{\underset{r=0}{\Sigma}} (r + 1)(r + 2) = 3{,}306\tfrac{2}{3} + \tfrac{1}{2}(2 + 462) + \tfrac{1}{12}(43 - 3)$$
$$= 3{,}542$$

(ii) To transform the series from the third term it is first noted that

$$\sum_{r=0}^{20} (r + 1)(r + 2) = 1.2 + 2.3 + \sum_{r=0}^{18} (r + 3)(r + 4)$$

i.e. $\sum_{r=0}^{20} (r + 1)(r + 2) = 8 + \sum_{r=0}^{18} (r + 3)(r + 4)$

Thus here the sum required is $\sum_{r=0}^{18} f(r)$ where $f(r) = (r + 3)(r + 4)$.

Thus $f(r) = r^2 + 7r + 12$

so that $\int_0^{18} f(r) \, dr = \left[\frac{r^3}{3} + \frac{7r^2}{2} + 12r \right]_0^{18}$

which gives $\int_0^{18} f(r) \, dr = 3{,}294$

Also since $f(r) = r^2 + 7r + 12$, $f_0 = 12$ and $f_{18} = 462$.
$f'(r) = 2r + 7$ and so $f_0' = 7$ and $f_{18}' = 43$. $f''(r) = 2$ so that $f'''(r)$ and higher derivatives are all zero.

Thus $\sum_0^{18} (r + 3)(r + 4) = 3{,}294 + \frac{1}{2}(12 + 462) + \frac{1}{12}(43 - 7)$

$$= 3{,}534$$

so that $\sum_0^{20} (r + 1)(r + 2) = 8 + 3{,}534$

$$= 3{,}542.$$

8.4   SPECIAL METHODS

If neither of the above transformations can be used to transform a slowly convergent series then it is sometimes possible to devise a special method applicable to that series.

Worked example 1

The evaluation of π.
The series for the inverse tangent (Gregory's series) is:

$$\tan^{-1} x = x - \frac{x^3}{3} + \frac{x^5}{5} - \frac{x^7}{7} + \ldots \text{ for } |x| \leqslant 1.$$

Putting $x = 1$ gives $\frac{\pi}{4} = 1 - \frac{1}{3} + \frac{1}{5} - \frac{1}{7} + \ldots$ which is known as Leibniz's series. This series converges very slowly, but other more rapidly converging series may be found using

$$\tan^{-1} A + \tan^{-1} B = \tan^{-1} \left(\frac{A + B}{1 - AB} \right)$$

e.g. (i) (Euler's method—not the same as that above)

If $A = \frac{1}{2}$ and $B = \frac{1}{3}$ then $\dfrac{A + B}{1 - AB} = 1$

and thus $\tan^{-1} 1 = \tan^{-1} \frac{1}{2} + \tan^{-1} \frac{1}{3}$

i.e. $\dfrac{\pi}{4} = \tan^{-1} \frac{1}{2} + \tan^{-1} \frac{1}{3}$

$$= \left(\frac{1}{2} - \frac{1}{3.2^3} + \frac{1}{5.2^5} - \frac{1}{7.2^7} + \cdots \right) +$$

$$\left(\frac{1}{3} - \frac{1}{3.3^3} + \frac{1}{5.3^5} - \frac{1}{7.3^7} + \cdots \right)$$

These series for $\tan^{-1} \frac{1}{2}$ and $\tan^{-1} \frac{1}{3}$ converge more rapidly than that for $\tan^{-1} 1$.

(ii) (Vega's Method)

If $A = \frac{1}{3}$ and $B = \frac{1}{7}$ then $\dfrac{A + B}{1 - AB} = \frac{1}{2}$

and thus $\tan^{-1} \frac{1}{2} = \tan^{-1} \frac{1}{3} + \tan^{-1} \frac{1}{7}$

but from (i) $\tan^{-1} \frac{1}{2} = \dfrac{\pi}{4} - \tan^{-1} \frac{1}{3}$

thus $\dfrac{\pi}{4} - \tan^{-1} \frac{1}{3} = \tan^{-1} \frac{1}{3} + \tan^{-1} \frac{1}{7}$

so that $\dfrac{\pi}{4} = 2 \tan^{-1} \frac{1}{3} + \tan^{-1} \frac{1}{7}$

$$= 2 \left(\frac{1}{3} - \frac{1}{3.3^3} + \frac{1}{5.3^5} - \frac{1}{7.3^7} + \cdots \right) +$$

$$\left(\frac{1}{7} - \frac{1}{3.7^3} + \frac{1}{5.7^5} - \frac{1}{7.7^7} + \cdots \right)$$

Similarly other series with even faster convergence may be established.

Worked example 2

The calculation of logarithms*

The series for $\log_e (1 + x)$

$$\log_e (1 + x) = x - \tfrac{1}{2}x^2 + \tfrac{1}{3}x^3 - \tfrac{1}{4}x^4 + \cdots$$

is only convergent for $-1 < x \leqslant 1$, and convergence is very slow for values of x near to ± 1. Euler's method may be used to transform the series for these values, and the following transformation when $|x| > 1$.

* See *Courant*

Since $\log_e (1 + x) = x - \frac{1}{2}x^2 + \frac{1}{3}x^3 - \frac{1}{4}x^4 + \dots$

then $\log_e (1 - x) = - x - \frac{1}{2}x^2 - \frac{1}{3}x^3 - \frac{1}{4}x^4 \dots$

so that $\frac{1}{2} \log_e \left(\frac{1 + x}{1 - x} \right) = x + \frac{1}{3}x^3 + \frac{1}{5}x^5 + \dots$ for $|x| < 1$.

Putting $\dfrac{1 + x}{1 - x} = \dfrac{p^2}{p^2 - 1}$ gives $x = \dfrac{1}{2p^2 - 1}$. For $|x| < 1$ we require

$\left| \dfrac{1}{2p^2 - 1} \right| < 1$ which is true for $p > 1$.

Thus $\frac{1}{2} \log_e \left(\dfrac{1 + x}{1 - x} \right) = \frac{1}{2} \log_e \left(\dfrac{p^2}{p^2 - 1} \right)$

$$= \tfrac{1}{2} \log_e p^2 - \tfrac{1}{2} \log_e (p^2 - 1)$$

$$= \log_e p - \tfrac{1}{2} \log_e (p - 1) - \tfrac{1}{2} \log_e (p + 1)$$

Hence $\log_e p - \tfrac{1}{2} \log_e (p - 1) - \tfrac{1}{2} \log_e (p + 1) = \dfrac{1}{2p^2 - 1} +$

$$\dfrac{1}{3(2p^2 - 1)^3} + \dfrac{1}{5(2p^2 - 1)^5} + \dots$$

i.e. $\log_e p = \tfrac{1}{2} \log_e (p - 1) + \tfrac{1}{2} \log_e (p + 1) + \dfrac{1}{2p^2 - 1} +$

$$\dfrac{1}{3(2p^2 - 1)^3} + \dfrac{1}{5(2p^2 - 1)^5} + \dots$$

$$\text{for } p > 1.$$

The logarithm of any rational number may be expressed in terms of the logarithms of prime numbers, e.g. $\log \frac{4}{5} = 2 \log 2 - \log 3$ and

$$\log 3 \cdot 85 = \log (77/20) = \log 7 + \log 11 - 2 \log 2 - \log 5.$$

Now if p is a prime number then $p - 1$ and $p + 1$ will both be even numbers and so have prime factors all smaller than p, so that $\log (p - 1)$ and $\log (p + 1)$ can be expressed in terms of the logarithms of prime numbers which are smaller than p.

e.g. If $p = 13$ then $p - 1 = 12$ and $p + 1 = 14$ so that

$$\log (p - 1) = 2 \log 2 + \log 3 \text{ and } \log (p + 1) = \log 2 + \log 7.$$

If $\log 2$ is known then this series can be used to find $\log 3$, but when $\log 3$ has been found $\log 5$ may be evaluated, etc., so building up the logarithms of the prime numbers, from which the logarithm of any rational number may be found.

Thus it is only necessary to evaluate this series for integral values of $p \geqslant 3$, and for these values this series converges very rapidly.

The value of log 2 may be found by transforming the series

$$\log 2 = 1 - \tfrac{1}{2} + \tfrac{1}{3} - \tfrac{1}{4} + \dots$$

by Euler's method for alternating series, or by noting that $\log 2 = \log 1 - \log \tfrac{1}{2}$ which is $-\log (1 - \tfrac{1}{2}) = \tfrac{1}{2} + \tfrac{1}{2}(\tfrac{1}{2})^2 + \tfrac{1}{3}(\tfrac{1}{2})^3 + \dots$ which converges quite rapidly.

8.5 EXAMPLES

1. Calculate log 2 correct to 3D using Euler's method for alternating series. (Tabulate $u_r = 1/(r + 1)$ for $r = 0(1)9$ to 4D and transform the series from the third term).

2. Show that the harmonic series $1 + \tfrac{1}{2} + \tfrac{1}{3} + \tfrac{1}{4} + \dots$ is divergent by transforming it using the Euler–Maclaurin integration formula.

3. Calculate log 3 correct to 5D given that log 2 = 0·693 147.

4. Evaluate the series $x - \tfrac{1}{2}x^2 + \tfrac{1}{3}x^3 - \tfrac{1}{4}x^4 + \dots$ at $x = 0.94$ correct to 3D.

5. Evaluate $\displaystyle\sum_{n=1}^{\infty} \frac{(-1)^{n+1}}{\sqrt{n}}$ correct to 3D.

6. Evaluate $\displaystyle\sum_{r=0}^{\infty} (\tfrac{5}{6})^r$ correct to 4D using the Euler–Maclaurin integration formula. Check your result by using the fact that this is a geometric series.

Answers to Examples

Chapter 1

§ **1.7** page 16

1 (a)
$$\Delta^n f_0 = f_n - n f_{n-1} + \frac{n(n-1)}{2!} f_{n-2} + \ldots + (-1)^n f_0$$

(b) $\Delta^3 f_0 = 0.006$

3 $f(x) \equiv x^4 - x - 21$ **4** $e^{3.5023} = 33.191\ 70$ (5D)

5 $0.783\ 172$ (6D)   **7** $0.419\ 320$ (6D)

§ **1.10.3** page 27

1 20.243	**2** -3.4518	**3** 2.5021
4 3.1942	**5** $-0.601\ 30$	**6** $2.528\ 274$

§ **1.16** page 41

1 $f(0.25) = 2.328$ **2** $f(0.543) = 6.303$

3 $f(2.53) = 19.786$ **4** $\operatorname{Sin} 1.234 = 0.943\ 82$ (5D)

5 $\sin (0.197) = 0.195\ 728$ **7** $\sec (1.263) = 3.300\ 77$

8 (ii) $0.442\ 755$ **9** $\tanh (2.217) = 0.976\ 543$

10 (i) $0.464\ 799$ (ii) $0.464\ 800$

11 $f(2.13) = 56.292\ (f(x) \equiv x^4 + 3.1x^3 + 2.7x)$

12 $f(0.137) = 1.177\ 11$ **13** $f(0.2427) = 1.120\ 357$

14 $15.734\ 22$ (5D)   **15** 25.4878 **16** 60.854 (3D)

Chapter 2

§ **2.5** page 53

1 $x = 1.365$ **2** $\begin{cases} \text{(i)}\ 0 < p < 1 \text{ we must take the positive root.} \\ \text{(ii)}\ x = 1.3327 \end{cases}$

3 $x = 1.057\ 02$ **4** $x = 1.863$ **5** $x = 0.767$

6 $x = 0.5630$ (4D)   **7** $x = 1.5570$ (4D)   **8** $x = -1.634$

§ **2.8** page 59

1 $x = 13.9702$ **2** $\begin{cases} \text{(i)}\ \pm 1 \times 10^{-6} \\ \text{(ii)}\ x_p = 1.661\ 30 \end{cases}$

3 $x = -2.6432$ **4** $x = -1.539\ 61$

5 $\begin{cases} \text{(i)}\ 25.487\ 81 \\ \text{(ii)}\ 3.4012\ (4D) \end{cases}$ **6** $\begin{cases} x = 1.523\ 61\ (6S) \\ x = 0.202\ 785\ (6S) \end{cases}$

7 $a = 1, b = 0, c = -10, d = 14$; real root $x = -3\cdot711\ 124$

8 $x = 1\cdot801$ radians

9 $x = -1\cdot3354$ **10** $x = 1\cdot199\ 679$

11 $x = 1\cdot147\ 753$ (7S)   **12** $x = 0\cdot309\ 017$ (6D)

13 $\begin{cases} \int f(0\cdot4) = 3\cdot8416 \\ x = 0\cdot778 \text{ (3D)} \end{cases}$ **14** $x = 1\cdot7230$ (5S)

15 $-0\cdot536\ 61$ **16** $\begin{cases} \rho = 0\cdot7 & x = 0\cdot1842 \\ \rho = 0\cdot9 & x = 0\cdot2495 \end{cases}$

17 $x = -0\cdot244\ 84$ (5D)

Chapter 3

§ **3.5** page 67

1 $L_1(x) = \dfrac{x^2 - 9x + 18}{10}$; $L_2(x) = \dfrac{x^2 - 7x + 6}{-6}$; $L_3(x) = \dfrac{x^2 - 4x + 3}{15}$

2 $L_0(x) = \dfrac{x^2 - 3x + 2}{12}$; $L_1(x) = \dfrac{x^2 - 4}{-3}$; $L_2(x) = \dfrac{x^2 + x - 2}{4}$

3 $L_0(x) = \dfrac{x^3 - 7x^2 + 10x}{-120}$; $L_1(x) = \dfrac{x^3 - 4x^2 - 11x + 30}{30}$

$L_2(x) = \dfrac{x^3 - 2x^2 - 15x}{-30}$; $L_3(x) = \dfrac{x^3 + x^2 - 6x}{120}$

4 (a) $L(x) \equiv \tfrac{9}{2}x^2 - \tfrac{13}{2}x - 3$;   (b) $f(2) = 2$

5 (a) $L(x) \equiv 3\cdot1x^2 - 1\cdot65$; (b) $f(0\cdot5) = -0\cdot875$

6 $f(x) \equiv x^2 - 1\cdot25$

§ **3.9** page 81

1 (i) $L(x) \equiv 3x^3 - 2x^2 + x + 2$ (ii) $f(1\cdot5) = 9\cdot125$

2 $L(x) \equiv x^3 - 2x - 3$ **3** $f(1\cdot235) = 3\cdot0356$ (4D)

5 $f(13) = 285\ 65$ **6** $L(x) \equiv x^4 + 4$

7 $f(6) = 586\cdot8$ (4S)

8 $f(0) = 5$. This result is exact because the data given is exact and function values are derived from a cubic function.

9 $f(10\cdot79) = 0\cdot0001$ **11** (i) $0\cdot219\ 332$; (ii) $665\ 23$; $768\ 20$.

12 (i) $f(0) = 10$; $f(2\cdot6) = 17\cdot248$, (ii) $L(x) \equiv 2x^3 - 5x^2 - 11x + 10$

13 $f(0) = 1\cdot1199$

14 $\begin{cases} x = 3 \\ y = -0\cdot596 \end{cases}$; $\begin{cases} x = 5 \\ y = -0\cdot761 \end{cases}$; $\begin{cases} x = 7 \\ y = -0\cdot792 \end{cases}$

Chapter 4

§ **4.6** page 99

1 $7\cdot153\ 78$ **2** (c) About 10   **3** $0\cdot211\ 51, 0\cdot426\ 25, 0\cdot644\ 11,$ $0\cdot865\ 00, 1\cdot088\ 84.\ 1\cdot315\ 53$

4 $V = 2\cdot089\ 81, x = 0\cdot314\ 49$ **5** $1\cdot4675$

6 $2\cdot4122$ **7** $0\cdot3616$ **8** $710\cdot858$

9 $-1\cdot756\ 23$ **10** $0\cdot064\ 661$ **11** $0\cdot831$

12 $1\cdot634\ 094\ 1$ **13** $0\cdot093\ 093\ 4$ **14** $0\cdot063\ 15$

Chapter 5

§ **5.7** page 114

1 -0.824, -2.47

4 (b) $0.820\,8$, -2

6 4.141

10 263.3, 2110

12 Error $= 0.0012$ in $f(1.3)$

13 $0.977\,92$

2 (b) $-0.955\,34$, $-0.877\,58$
  (c) $0.435\,0$ $0.389\,4$

5 0.0399

8 32.52

11 -14.793

14 (a) (b) (c) 5S, (d) 4S

Chapter 6

§ **6.2.4** page 127

1 x 0   0·1   0·2   0·3   0·4   0·5
  y -1 -1.1097 -1.2381 -1.3848 -1.5507 -1.7379

2 x 0   0·1   0·2   0·3   0·4   0·5
  y 0·5   0·498   0·492   0·486   0·487   0·502

3 x 0   0·1   0·2   0·3   0·4   0·5   0·6
  y 0 -0.084 -0.134 -0.143 -0.109 -0.029 $+0.094$

4 x 0   0·1   0·2   0·3   0·4   0·5
  y 0·5   0·5944   0·6751   0·7394   0·7850   0·8109

§ **6.2.6** page 130

1 $y = 4x + x^2 + x^4 + \frac{1}{5}x^5$ **2** $y = -1 - x^3 - \frac{1}{2}x^6$

3 $y = 1 - \frac{1}{24}x^4 + \frac{1}{120}x^6$

4 $y = 1 - x + x^2 - \frac{2}{3}x^3 + \frac{5}{6}x^4 - \frac{4}{5}x^5 + \frac{3}{2}x^6$

5 $y = -\frac{1}{3}x^2 - \frac{2}{9}x^3 - \frac{2}{9}x^4 - \frac{8}{27}x^5 - \frac{40}{81}x^6$

§ **6.2.8** page 134

1 x 0   0·05   0·10   0·15   0·20   0·25   0·30
  y 0   0·0001   0·0010   0·0034   0·0079   0·0154   0·0265

2 x 0   0·05   0·10   0·15   0·20   0·25   0·30
  y 0 -0.046 -0.084 -0.114 -0.134 -0.144 -0.143

3 x 0   0·1   0·2   0·3   0·4   0·5
  y -1 -1.1097 -1.2381 -1.3848 -1.5507 -1.7379

4 x 0   0·1   0·2   0·3   0·4   0·5   0·6   0·7   0·8   0·9   1·0
  y 0·1   0·100   0·103   0·110   0·123   0·145   0·178   0·226   0·292   0·384   0·512

5 x 0   0·05   0·10   0·15   0·20   0·25   0·30   0·35   0·40
  y 0·25   0·2500   0·2501   0·2503   0·2507   0·2513   0·2523   0·2536   0·2554

§ **6.3.2** page 138

1 x 0   0·1   0·2   0·3   0·4   0·5
  y 1   0·900   0·804   0·712   0·626   0·548

2 x 0   0·1   0·2   0·3   0·4   0·5
  y 0   0·100   0·200   0·301   0·404   0·511

3 x 0   0·1   0·2   0·3   0·4   0·5
  y 1   0·9002   0·8010   0·7028   0·6057   0·5095

4 x 0   0·05   0·10   0·15   0·20   0·25   0·30
  y 0   0·0551   0·1208   0·1980   0·2876   0·3911   0·5100

§ **6.3.4** page 140

1, 2, 3 Compare results with those in Ex. 6.3.2, 1, 2 and 3 respectively.

4 $\ddot{x} = \cos \dfrac{t}{2}$

Chapter 7

§ **7.2.6** page 148

1 $y = 2{\cdot}403 + 0{\cdot}235x$ **2** $y = 21{\cdot}80 + 0{\cdot}09929x$

3 $y = 12{\cdot}73 - 0{\cdot}3181x$ **4** $y = 0{\cdot}439 - 0{\cdot}033x$

5 $a = 2{\cdot}01,\ c = 2{\cdot}98$ to 2D   **6** $y = 0{\cdot}494 + 0{\cdot}272x$ to 3D

7 $a = 4{\cdot}325,\ n = 0{\cdot}426$ to 3D

§ **7.3.1** page 152

1 $y = -0{\cdot}04762 + 1{\cdot}392x + 0{\cdot}5833x^2$ to 4S   **2** $y = 50{\cdot}78 + 16{\cdot}36x - 14{\cdot}76x^2$ to 4S

3 $y = 101{\cdot}3 + 1{\cdot}424x - 0{\cdot}06142x^2$ to 4S

4 $h = 0{\cdot}454 + 6{\cdot}86t - 0{\cdot}451t^2$ to 3S   **5** $y = 7{\cdot}33 - 19{\cdot}30x + 13{\cdot}71x^2$ to 2D

6 $y = 100 + x - 2x^2$

§ **7.4.1** page 154

1 $a = 2{\cdot}02,\ b = 3{\cdot}01$ **2** $a = 2{\cdot}49,\ b = 1{\cdot}53,\ y \simeq 2{\cdot}54$

Chapter 8

§ **8.5** page 161

1 $0{\cdot}693$ **3** $1{\cdot}098\ 61$

4 $0{\cdot}663$ **5** $0{\cdot}605$

6 $6{\cdot}0000$

Bibliography

ALLEN, D. N. DE G. (1954) *Relaxation Methods in Engineering and Science.* McGraw-Hill, New York and Maidenhead.

British Standard Letter Symbols, Signs and Abbreviations. BS 1991, Part 1 (1954). British Standards Institution, London.

British Standard Flow Chart Symbols. BS 4058, Part I (1966). British Standards Institution, London.

BUTLER, R. and KERR, E. (1967) *An Introduction to Numerical Methods.* Pitman, London.

COMRIE, L. J. (1966) *Chambers Shorter Six Figure Tables.*

COURANT, R. (1937) *Differential and Integral Calculus.* 2nd edn. Blackie, Glasgow.

FOX, L. (1957) *The Numerical Solution of Two-Point Boundary Problems in Ordinary Differential Equations.* Pergamon, Oxford.

HARTREE, D. R. (1958) *Numerical Analysis.* 2nd edn. Oxford University Press.

Interpolation and Allied Tables (1955). H.M.S.O.

LIPKA, *Graphical and Mechanical Computation.* Wiley, New York and London

REDISH, K. A. (1961) *Introduction to Computational Methods.* English University Press.

WHITTAKER, SIR E. and ROBINSON, G. (1944) *Calculus of Observations,* 4th edn. Blackie, Glasgow.

Index

166